THE ASTRONOMY OF SOUTHERN AFRICA

THE ASTRONOMY OF SOUTHERN AFRICA

PATRICK MOORE

AND

PETE COLLINS

LONDON

ROBERT HALE & COMPANY

First published in Great Britain 1977

ISBN 0 7091 6176 X

ROBERT HALE & COMPANY
Clerkenwell House
45-47 Clerkenwell Green
London
Ecir Oht

Printed in the Republic of South Africa
by Citadel Press, Lansdowne, Cape

TO

MOYA

(who else?)

CONTENTS

Acknowledgements 13

Introduction 15

1. Prologue: The Stars in Folklore 19
2. The Cape: Early Days 21
3. The Cape: Years of Progress 44
4. The Cape: From Stone to Sutherland 71
5. The Republic Observatory: Triumph and Controversy 92
6. The Leiden Station 107
7. The Yale Observatory, Johannesburg 111
8. The Radcliffe Observatory: End of a Story 114
9. Boyden: The Observatory on the Hill 120
10. The Lamont-Hussey Observatory 130
11. Thc Natal Observatory: Promise and Failure 132
12. Into Deep Space: The Great Dish at Hartebeeshoek 137
13. Temporary Observatories 142
14. Astronomy in the Universities 144
15. The Amateurs 149

Index 155

LIST OF ILLUSTRATIONS

Pete Collins and Patrick Moore on the set of The Southern Sky television series 16
Father Guy Tachard's Observatory, Cape of Good Hope, 1685 22
Peter Kolben (Kolbe or Colbe), the first official Cape Astronomer 27
The Abbé Nicolas Louis de la Caille 28
A page of La Caille's notebook 30
Dr Patrick Moore reading La Caille's notebook 31
Letter from the Prince of Orange and Nassau 34
Old houses in Strand Street, Cape Town 35
Memorial plaque, designed by Sir Herbert Baker 36
La Caille's Constellations 38
La Caille's sundial in the courtyard of number 7, Strand Street 40
La Caille's map of his triangulation 41
Title page of La Caille's *Astronomiæ Fundamenta* 42
Henderson's "Dismal Swamp" in 1911 46
Sir Thomas Maclear ("A good Irish face") 48
Sir John Herschel 50
Photographic reproduction of a camera lucida sketch by Herschel 52
A highly-finished camera lucida sketch by Sir John of Devil's Peak 53
Camera lucida sketch by Sir John of his daughters 54
The first page of Sir George Airy's letter of instructions to Maclear 56
Measurement of the base line on the Grand Parade in Cape Town 58
View from La Caille's northern point 60
Another view from La Caille's northern point 61
The station on Sneeukop, Hottentots-Holland 62
The Herschel Monument as it originally appeared in the grounds of Feldhausen 64
The inscription on the Herschel Monument 65
The Herschel Monument during building operations 66
On board the Windsor on the Herschels' homeward voyage 67
The Royal Observatory, Cape Town, in 1865 68
Transit instrument at the Cape Observatory 1852 69
Sir David Gill 72
The Comet of 1882 as seen in South Africa 75
Gill's historic photograph of the 1882 comet 76
Sketches from *The Graphic*, August 27th, 1892 77

The Victoria Telescope, Royal Observatory, Cape of Good Hope 78
Staff of Royal Observatory, Cape of Good Hope, 1906 80
Halley's Comet as seen from Port Elizabeth 81
The Main Building at the Cape Observatory 84
The Old Library at the Cape Observatory 85
The telescope domes at Sutherland Observatory 86
Sir Richard van der Riet Woolley at Sutherland 87
The Official Opening of the Sutherland Observatory, 15 March, 1973 88
A scene at the Official Opening of the Sutherland Observatory 89
The 74-inch Telescope Building at Sutherland 90
The Invitation to the Opening of the Government Meteorological Observatory 93
Formal Opening of the Government Meteorological Observatory 94
The Government Meteorological Observatory in Johannesburg in 1905 95
Robert Thorburn Ayton Innes. The first Union Astronomer 96
Installation of the 26½-inch telescope at the Union Observatory, Johannesburg 97
The Franklin-Adams Telescope at the Union Observatory 98
Interior of observing hut of Franklin-Adams Telescope, Union Observatory 99
The four successive Union Astronomers 100
The Old Government Meteorological Observatory, Johannesburg 101
The Republic Observatory, Johannesburg 103
The Rockefeller 16-inch twin astrograph at Leiden Southern Station 108
Dr Theo Walraven demonstrating the Rockefeller 16-inch twin astrograph 109
The 36-inch Flux-collector telescope at Leiden Southern Station 110
The Old Yale Observatory building 112
"Canopsus" Street 116
The Radcliffe Observatory, Pretoria, in 1952 117
The Radcliffe Observatory, Pretoria, in 1975 118
An aerial view of Boyden Observatory 121
The 60-inch Rockefeller Telescope at Boyden 123
Professor Alan H. Jarrett, and Patrick Moore inspect the Metcalf Telescope 126
The Lamont-Hussey Observatory on Naval Hill, Bloemfontein 131
Edmund Neison 132
Lunar sketches by Edmund Neison of Natal 134
The 85-ft (26 metre) Great Dish at Hartebeeshoek 139
Steve Heninger and Patrick Moore beside the Laser-Transmitter 143
Professor Brian Warner explains a dwarf nova outburst 145

The Planetarium at the University of the Witwatersrand 147
The Hoba West meteorite 148
Dr Patrick Moore and Jack Bennett 150
Chris Papadopoulos, star-atlas compiler 151
David Block in his observatory at Krugersdorp 153

ACKNOWLEDGEMENTS

During the compilation of this book we have had the most generous help from many people, particularly:

Dr W. Ahlers
Jack Bennett
Miss Joan Biddles
David Block
Don Briscoe
Dr Hamish Campbell
Mrs Anna Cunningham
Dr W. S. Finsen
Thomas Geary
Andrew Gray
Steve Heninger
Dr A. H. Jarrett
Robin Knox-Grant
J. D. Laing
L. D. Meyer
Professor L. C. Nicolaysen
Dr George Nicolson
Danie Overbeek
Dr Jan Schutte
Fred Spence
David Stevenson
Duncan Stepto
R. Vice
Dr T. Walraven
Professor Brian Warner
Sir Richard Woolley

To all these – our most grateful thanks

PATRICK MOORE
PETE COLLINS

INTRODUCTION

THE BOOK you are about to read is not the book we originally intended to write. Neither does it have the same title.

While working on the television series *The Southern Sky*, it seemed to us that people might be interested in reading how the programmes were produced – how the idea of such an unusual series originated, and in the day-to-day experiences, good and bad, which we encountered in the course of production. We also felt that by setting it all down in black and white we could ourselves recall, in after years, what had proved to be a highly pleasurable and rewarding period in the lives of us both. And yet things simply did not work out in that way.

First, there was the question of the theme. To write a book about the astronomy of South Africa it was, of course, necessary to find out all the details about it; how it all started, how it developed – and why. This led to a considerable amount of correspondence with people who could supply us with essential information. It also lead to the reading of many books and papers, to visits to the most unexpected places, and to the spending of hour after hour in such repositories of knowledge as the third basement in the Old Library of the University of the Witwatersrand.

The more we delved, the more obvious it became to us that to produce a book on the lines we had originally intended would satisfy neither ourselves nor those who read it. Even though we were restricted by considerations of modern publication costs, it was essential to set down at least the broad outlines of the great South African astronomical story, so that both those who live in the Republic and those who live abroad would realize that the story is a proud one indeed – and one which has had a profound effect upon the science in all other parts of the world.

Next, there was the matter of the title. At first there was no doubt in our minds, and we went back straight to Sir John Herschel, an engaging character about whom much will be said later on. He arrived in the Cape on 15 January 1834, set up his great telescope of 20 feet focal length, and proceeded to study the southern stars. He also kept a diary which has been described as personal, scientific and peculiar. Despite the appalling handwriting in which it was set down, it has been lovingly edited by Dr David S. Evans, Terence J. Deeming, Betty Hall Evans and Stephen Goldfarb, a team highly qualified to undertake such an important task. It covers Herschel's entire Cape period, from 1834 to 1838, and it is by far the best bedside book that anyone with the slightest interest in astronomy could have. To say that Herschel the man "comes alive" in its pages is woefully understating the case. His family, his servants, and all who were in any way connected with him also come alive in the most startling fashion. So do the famous people who visited him – such as Charles Darwin, who had grave fault to find with the astronomer's personal manners!

Herschel was highly sociable, and he faithfully recorded the visits of his friends and acquaintances to the observatory and house. Of course, the presence of a giant telescope on the premises was a highly unusual circumstance, and few visitors failed to delay their departure until night had fallen, in the hope – which was seldom disappointed – that they might be invited to "have a peep". Herschel records these impromptu observations in

Pete Collins and Patrick Moore on the set of The Southern Sky television series in the SABC-TV studios at Auckland Park.

Photograph – Don Briscoe

many ways. "Mr P. Smyth came to sweep through the Great Magellan . . ." "In evening S. came. Shewed him α Centauri - Saturn & my blue Planetary nebula near the Cross . . ." "Mrs S. and family to *tea & stars* . . ." "Lieut. Williams came to *tea & stars* bringing over Wm. Petrie junr. . . ." The delightful phrase *tea & stars*, usually underlined, occurs in several places in the diary. It conjures up the mental picture of Lady Herschel presiding over a silver tea-service while Sir John hands round thinly-sliced bread and butter, their guests perched on the edges of their chairs in eager anticipation of the promised introduction to the glories of the clear South African heavens, via the great 20-ft. telescope.

So what better title for a book on southern hemisphere astronomy could there be than *Tea & Stars*, in suitably graceful – and grateful – acknowledgement of the kindly hospitality which the television unit and we ourselves received from the distinguished astronomers all over the Republic?

Yet . . . in view of the changed nature of our approach, was it now really apt? Regretfully, we were inclined to think not. It was imbued with a faint air of dilettantism, foreign indeed to the record of hard and often gruelling toil that was being steadily uncovered during our deepening research. So *Tea & Stars* was dropped, with a rueful mental shrug to the wraith of Sir John.

But this is the place to record, quite clearly and without reservations of any kind, our gratitude to the many busy men and women who paused in the course of their work to supply us with essential information. We are keenly aware of what such disruption means to scientists who are engaged in progressive research. Yet not once, either in the making of the television series or in the gathering of material for this book, was a voice raised in complaint.

In fact, the book you are about to read had its origin in television more than two decades ago. The writers have known each other for a great many years, and in 1952, which was long before lunar travel was regarded as a practical possibility, a film which was produced which described a Moon voyage with surprising accuracy. The producer/director was Pete Collins. The technical adviser was Patrick Moore, whose subsequent BBC television series, *The Sky at Night*, has continued monthly, without a break, since April 1957. It was logical that when television came to South Africa a television series should be presented here too. *The Southern Sky* was the result, again produced and controlled by Pete Collins and with Patrick Moore as presenter.

It is fair to say that the series, which ran throughout the first half of 1976, aroused a great deal of interest. (Among those who appeared in it were Sir Richard Woolley, Dr Alan Jarrett, Professor Brian Warner and Dr T. Walraven.) It also led to a great many inquiries about the story of South African astronomy. We were asked to write this book; we have done so, and we hope that it will be acceptable both to those who take a passing interest in astronomy and to those who are seriously interested in the history of science in general, South African science in particular.

If we succeed in these aims, we will be well satisfied.

PATRICK MOORE
PETE COLLINS

Sandown, Transvaal
June 1976

1. PROLOGUE: THE STARS IN FOLKLORE

ASTRONOMY is the oldest science in the world. Even the cavemen of so long ago must have been star-gazers, even though they had no idea of how the universe is made up. As human intelligence developed, astronomy developed too, and the stars were divided up into constellations, while the planets were named after the ancient gods and goddesses. Legends came from many lands, and it has been said that the sky is a vast picture-book. The constellations include such characters as Hercules, the great hero; Perseus, who killed the hideous sea-monster by showing it the Gorgon's head and turning it to stone; and Orion, the hunter. There is, too, the ship Argo, which carried Jason and his companions in their quest of the Golden Fleece.

Most people are familiar with the Greek legends, but almost every other continent has its own folklore, and Africa is no exception. Unfortunately the African legends are not well documented, and there seems to have been little in the way of true astronomy before the arrival of the first Europeans. Yet the record is not entirely blank. There is evidence that since very early times the Africans have known that the Earth is in motion relative to the Sun, and there is an old song which, when translated, means "I shall worship you and go round you, as the Earth worships the Sun". Cave engravings show the five naked-eye planets revolving round the Sun, and Venus was the object of special attention, since it was believed that any child born early in the morning when Venus was visible would be beautiful – and would have many lovers. One such child was Nandi, mother of the famous Zulu leader Shaka. Mars, needless to say, was thought to have a baleful influence; nothing seems to have been recorded about Mercury, Jupiter or Saturn.

In African mythology the Earth was a bowl made of clay and rocks, criss-crossed by mountains and rivers; it was carried on the back of a giant tortoise, which floated round the Sun in a huge, blue ocean called the Sea of Eternity. It was believed that when the tortoise tips to one side, the blue surface can be seen as well as the Sun, and the solar rays shine into the bowl, causing daylight; when the tortoise floats upright the edge of the bowl acts as a shield against the Sun, revealing nothing but the blackness of Heaven.

From this, it seems that the Africans had at least some kind of cosmogony, and, inevitably, there was a great deal of both sun-worship and moon-worship. To the Pygmies the supreme god was Khonvum, who controls all the phenomena of the sky. When the Sun dies at night, Khonvum collects the broken pieces of stars in his sack and throws armfuls of them at the Sun, mending it so that it can rise unharmed next morning. In a legend from the Zambezi, the Moon "used to be very pale, and did not shine; she was jealous of the glittering Sun. Once, when the Sun was looking at the other side of the Earth, the Moon stole some of his fire to adorn herself. In anger, the Sun splashed the Moon with mud, and the dark patches are still to be seen. But the Moon was bent on vengeance, and whenever she can catch the Sun off his guard she spatters him with mud in his turn. Then the Sun stops shining for some hours, and the whole Earth is in terror." Surely this can only be a reference to solar eclipses.

The Moon was important in Zulu lore because it was thought to influence animals and plants. For instance, it was held that peas planted between two and three days before full moon yielded more than twice the crop of those planted earlier. No woman "at the time of her moon" was allowed to eat amasi (curdled milk) or to become angry – or to enter a cattle-kraal, or to sit on the wooden stools normally used by the men. Among the Bushmen, the phases of the Moon were explained quite simply; the Moon regularly offends the Sun, and is whittled away before crying for mercy and being gradually restored. The Bushmen known as the Ba-letsatsi, or Men of the Sun, were disinclined to work on a day when the Sun rose behind clouds, on the pretext that it might damage their hearts. It has also been said that the Bushmen held their ceremonial dances at the times of new moon and full moon. Dancing begins with new moon, because the dark nights have ended; it is continued at full, so that those taking part can enjoy the delightful coolness after the heat of the day.

In some African paintings the Sun is shown as a circle with a cross through it – in fact a swastika, the oldest symbol of light in Africa. The full moon is a circle with a central dot, indicating Love; the crescent moon indicates Tragedy. Moreover, there are records of the Three Healers (Orion's Belt), the Ploughing Stars (the Pleiades), the Starry River (the Milky Way) and the Tree of Life (Crux Australis, the Southern Cross). Crux is the holiest of all symbols, put into the sky specifically to act as a guide to wanderers who would otherwise lose their way.

Folklore of this kind is reasonably well authenticated, but it cannot be said to be rich, and there was no overall pattern comparable with that in many other countries. So apart from some casual records of observations of comets, it seems that the astronomy of South Africa has a definite beginning – in the year 1685, when Father Guy Tachard landed at the Cape.

2. THE CAPE: EARLY DAYS

"ASTRONOMY has always been strong in South Africa – for more than a hundred and fifty years."

These words were spoken by Sir Richard Woolley, former Astronomer Royal of England, now Director of the South African Astronomical Observatory, in a television interview with Patrick Moore at the great new installation at Sutherland in the Karoo. The statement may have come as a surprise to many people, even those who take an interest in the Republic's colourful history. A century and a half ago the start of the Great Boer Trek to the North still lay eight years in the future, and in the Eastern Cape the latest contingent of British settlers had only had six years in which to size up their new home. Yet it was in the very year of their arrival – 1820 – that an Order in Council was issued in Great Britain which resulted in the establishment of the Royal Observatory at the Cape. This shows how important astronomy was considered to be to the welfare of the colony.

It was originally assumed that the main work at the Cape would centre on navigational astronomy, and the Observatory was therefore erected within sight of Table Bay, so that time signals could be given to ships. The fact that the selected semi-swamp site was infested with snakes, leopard and hippopotami was considered to be of secondary importance, both by those who selected it and by most of the early astronomers themselves. From their notes and letters, one has the impression that had they not had these typically African fauna to contend with they would have been quite disappointed, particularly after so long a journey.

Of course, it was quite correct to say that the officially-organized astronomy of South Africa began with the founding of the Cape Observatory, but there had been some earlier astronomers whose labours cannot be glossed over. In the Monthly Notes of the Astronomical Society of South Africa for June 1948, Sir Richard gave an historical account of the work of some of these men which has proved to be of great value to later historians, including the present writers. From these notes, and others of a similar nature, it is found that the earliest recorded astronomical observations from the area date back as far as 600 B.C., when, as recorded by Herodotus (iv. 42), Phoenician sailors rounded the Cape on their way to the Far East.

There followed a long lull, which lasted until the voyages of the great Portuguese navigators of the fifteenth century. However, some records of comets were made, and it is worth mentioning the case of the comet of 1664, which was used by the then Governor of the Cape as a kind of threat against sailors, soldiers and general workmen who tended to use Sunday as a day for drinking and idling instead of going to church. There was no resident clergyman, and the rôle was filled more or less unofficially by one Ernestus Back, who proved to be one of the worst offenders; he was described as a man "addicted to intemperance and loose living". When the comet was brilliant, in February 1665, Back was at his very worst, and the celestial portent gave the authorities a good excuse to put him on a ship and dispatch him to Batavia. From all accounts, he never returned.

But better things lay ahead, and the main story may be said to begin with the advent of the redoubtable French Jesuit priest, Father Guy Tachard, who came ashore at the Cape in 1685 when he was on his way to Siam, and stayed long enough to establish an ob-

Father Guy Tachard's Observatory, Cape of Good Hope, 1685.

– Courtesy of Africana Museum, Johannesburg

servatory near the Castle. A picture of this observatory is shown here. It first appeared in a book called *Voyages de Siam des Pères Jesuites*, dated 1686; and it has been the subject of much pleasurable discussion and conjecture ever since.

That Father Tachard's observatory really looked like this is doubtful in the extreme. One imagines that when he returned home he described it to a brother Jesuit who had the taste for art – and that from the good Father's description and with his ultimate Siamese destination firmly in the artist's mind, the drawing of this delightful pagoda-like structure eventually emerged. It has even been suggested that the entire graphic concept owes much to the traditional Willow Pattern, which at that time was just becoming popular in France.

It is almost a pity to dissect the drawing and break it up into its component parts – a procedure, however, which historical accuracy demands.

Let us then first consider the jaunty little sailing ship which is heading out of the picture before a spanking breeze. In the earliest version of the drawing – and at least three, from various sources, exist – this ship does not appear at all, and its place is taken by a canoe of markedly West African aspect. The latter vessel gives the impression of being about to run up on to the grassy shore – a circumstance which, had it occurred in real life, would have occasioned the astronomers inside the observatory considerable dismay, persuading them that the cannibals they had heard so much about before leaving home had come for them at last.

The mountains which form the background are all wrong. They bear a faint resemblance to the Hottentots Hollands, but if that is what they are meant to be they shouldn't be there at all.

The Castle at Cape Town has come off rather better, and it is indeed quite recognizable. As the artist had almost certainly never set eyes on the original, this may be regarded as a major imaginative feat. He has, of course, placed it on the wrong side, but he probably thought that it looked better there.

The double row of trees to the left of the Observatory building is obviously meant to represent the gardens of the Dutch East India Company. Again, if this is so, they should not be there. Clearly, the intention was to locate Father Tachard's observatory on the shore side of the gardens. As this site was at that time solidly occupied by Fort Knocke, the gravest objections to its pictured situation immediately arise.

As to what it is now fashionable to call the picture's ambience, it has been suggested by an irreverent Afrikaner that the early colonists obviously enjoyed Divine favour; why else would a bountiful rain of *vet-koekies* be descending upon the land? There is, however, the alternative possibility that the large white blobs are meant to represent the stars in the South African sky, and this conclusion would seem to be borne out by the silvery moonlight in which the entire scene is bathed, but with this idea the critic had no patience.

Readers who live outside South Africa may be puzzled by the mention of the gardens of the Dutch East India Company at the Cape. Therefore, it may be as well at this point to sketch in the circumstances which led up to the foundation of the modern South African state. Anyone coming from Kimberley will certainly think that the country as we see it owes its being to the diamond; a citizen of Johannesburg will wave an arm to indicate the now grass-covered mine dumps, and state flatly that South Africa was founded on gold. Yet neither view is correct. Diamonds and gold were, naturally, of tremendous importance, but what really put South Africa on the map – literally – was some stuff which can now be bought in tubes at about a rand for a phial of fifty tablets. It is called Vitamin C.

Monuments have been erected to the Voortrekkers and to the 1820 Settlers. Jan van Riebeeck stares solemnly across Cape Town foreshore to a point which he assuredly never saw, as it was not there in his day (it was not reclaimed from the sea until a good three hundred years later). But the white crystalline powder, present in fresh greenstuffs, goes unhonoured by plaque or stone. It would admittedly be rather difficult to imagine what form such a monument could take (perhaps Henry Moore or even Rowland Emmet would provide some ideas!) but be this as it may, there would have been no need for a Dutch expedition to the Cape in 1652, to establish a staging-point between Europe and the Dutch East Indies, had not the power of Vitamin C been discovered. It was the one known remedy for that disease dreaded by all sailors: scurvy.

Almost Jan van Riebeeck's first act after the expedition had come ashore was to establish extensive and, from contemporary accounts, very beautiful gardens. These Company Gardens still exist in Cape Town, as a public park, and their extent is often a surprise to visitors. A Castle was also built – you can see it today – and a small settlement was founded outside its walls. The whole was surrounded by a mighty hedge of wild almonds, parts of which survive.

Digging gardens, building castles, founding settlements and planting hedges to go round them is all very well, provided that you can find them again next time your ship passes that way. Unfortunately, because of the rather vague nature of navigational astronomy at that time, it was by no means uncommon for a ship-master to find his craft bobbing about in the extremely rough waters of the South Atlantic, having missed the Cape altogether. This situation, at once perilous and costly, was clearly intolerable. Because of crude instrumentation, however, it had to be borne, although some of the country's best minds were already trying to find an astronomical solution to what was always called the 'longitude problem'.

The great Edmond Halley, of comet fame, was one such man. In 1676 he landed on the remote island of St Helena, set up an observatory, and proceeded to catalogue the positions of 341 southern stars. His correspondence fully covers his work there. The main trouble was the weather; St Helena has by no means an ideal climate. Halley did his best to chart the southern sky, which at that time was more or less unexplored, but if really valuable results were to be obtained it was clear that a more favourable site was needed.

After this digression, let us come back to Father Guy Tachard, who, as we have noted, arrived at the Cape in 1685. As a Jesuit priest, he could not know what sort of reception to expect from the Dutch authorities. He need not have worried. At that time the Cape had one of its best governors: Simon van der Stel, who assumed office in 1679 and remained for twenty peaceful years. The Dutch gave Tachard's party an excellent welcome, and throughout their stay did everything possible to be helpful. Tachard, for his part, was only too anxious to be courteous. Indeed, he may have been over-anxious, and at the outset a Gilbertian muddle occurred with regard to the firing of the correct number of gun salutes. To quote Tachard's own account, from his book *Voyage de Siam des Pères Jesuites:* –

"It was agreed upon that the Fort should render Gun for Gun when our Ship saluted it. This Article was ill explained, or ill understood, by these Gentlemen, for about ten of the Clock my Lord Ambassador having ordered seven Guns to be fired, the Admiral answered with only five Guns, and the Fort fired none at all. Immediately the Ambassador sent ashore again, and it was determined that the Admirals Salute should pass for nothing, and so the Fort fired seven Guns, the Admiral seven Guns, and the other Ships five, to salute the Kings Ship, which returned them their Salutes, for which the Fort and Ships gave their Thanks."

This little contretemps having been safely overcome, the travellers went ashore to a warm greeting. They landed their measuring instruments, and prepared for their main task: the determination of the exact longitude of the Cape, which was not then known with any accuracy at all. The method Tachard hoped to use involved the phenomena of the four principal satellites of Jupiter, which had been discovered by Galileo and his contemporaries in 1610 – the earliest days of telescopic astronomy. Accurate tables of the satellites had been worked out by the Director of the Paris Observatory, G. D. Cassini, and Tachard had brought them with him. If he could find the exact moment of one of the phenomena – an occultation of a satellite by Jupiter, the start or end of the transit of a satellite across the planet's disk, or the beginning or end of a satellite eclipse by Jupiter's shadow – he would also know the standard time; and comparing this with the local time would give him the longitude of the observing site. (Today, the site is occupied by the Old Supreme Court). So Tachard and his team set up their telescope, and on 2 June they started work. To quote once more:

"A Pendulum Clock with Seconds, made by Monsieur Thuret at Paris, being set at to* Hour, as near the true Hour as we could guess, not having as yet corrected the Clock, we began the following Observations.

"The first satellite** appeared in the Evening at eleven and three Minutes of the Clock not as yet corrected, distant from Jupiter, somewhat less than the Diameter of the same Jupiter.

"By the Telescope we saw two parallel Bands or Streaks upon the Body of Jupiter, the one larger towards the Southern Border, and the other narrower towards the Northern.

"The first Satellite began to touch the Border of Jupiter at 11 a Clock 37 M 30 S at eleven a Clock 58 M 5 S the Satellite was no more to be seen.

"These Observations were made with an Excellent Telescope twelve foot long, of the late Monsieur le Bas: the Hours are still of the Pendulum not rectified.

"We constantly observed Jupiter until two of the Clock five Minutes after Midnight, at what time he was hid behind the Lyon Mountain, which limited our Sight towards the West, so that we could not that day see the Emersion of the first Satellite."

This, of course, was a preliminary attempt, and Tachard was by no means satisfied with it. For the next ten days or so he made further measurements, and came to the final conclusion that the Cape was three degrees further west in longitude than had been previously thought. As a matter of fact his observations were not accurate enough to be really useful, but this was not Tachard's fault; he did his best with the inadequate equipment available to him. Moreover, he made some general observations of the southern stars – the first telescopic observations of them from South Africa:

"The foot of the Crozier† marked in Bayer is a Double Star, that is to say, consisting of two bright Stars distant from one another about their own Diameter only, much like to the most Northern of the Twins; not to speak of a third much less, which is also to be seen, but further from these two.

**Sic.*

**Tachard spells this word "satelles" throughout.

†The Southern Cross.

"Under the Croziers there are several Places of the Milky Way, which seem to be filled, if you look through the Telescope, with an infinite number of Stars.

"The two Clouds which are near the South Pole appear not to be a cloud of stars, like the Præsepe Cancra, nor yet a duskish Light, as the cloudy part of Andromeda: nothing hardly is to be seen there with great Glasses, though without a Glass one may see them very white, especially the great Cloud.

"No constellation in the Heavens looks so lovely as those of the Centaur and the Ship. There are no fair stars near the Pole, but a great many little ones. Bayer and other Books that speak of them, omit a great many, and most of those they set down appear not in the Heavens in the same Situation."

There are various interesting points here. The "foot of the Cross" is, of course, the celebrated double Alpha Crucis, which is one of the most spectacular pairs in the sky; it is lettered Alpha in the 1603 catalogue of Johann Bayer, who introduced the system of alloting the stars Greek letters. The "northern of the Twins" is Castor in Gemini, which was also known to be double; it was wide and easy in Tachard's day, though it is not so easy now, because it is a binary (revolution period 350 years) and is closing up. Tachard's description of the Magellanic Clouds is also very reasonable – and his telescope must have been at least adequate by seventeenth-century standards, as otherwise it would not have shown the two equatorial belts on Jupiter.

Clearly the Dutch authorities realized that they had skilled astronomers in their midst, and they spared no effort to make them comfortable. The Ambassador himself joined in the observing, and after he had gone "the Commissary with the Heers St Martin, Vanderstel and Bocheros stayed with us in the Observatory till ten a Clock at Night". When the time for departure came, cordial good-byes were said; the Dutch presented the astronomers with "Presents of Tea and Canary Wine", while Tachard left his hosts a microscope and a burning-glass. Moreover, the Jesuits were most impressed with the good spirit among all the inhabitants of the Cape, regardless of nationality, colour, status or religion. All in all it was a particularly pleasant episode. Tachard never returned to the Cape, but he must have carried away very happy memories of it and its people.

After the good Frenchman, Father Tachard, came the bad German Peter Kolbe, who – despite any character defects he may or may not have had – was sent to the Cape in 1705 as the first astronomer to be officially dispatched there for the express purpose of observing celestial phenomena. It is true that his patron, Baron Krosick, expected Kolbe to study the natural history of the Cape area as well as determining its geographical position. However, Baron Krosick also sent another astronomer northward to make observations complementary to Kolbe's for measuring the distance of the Moon, so that astronomical work was clearly intended to be at least one of Kolbe's chief tasks.

Quite possibly he was nothing like so bad as his contemporaries made out. The chief complaint against him seems to have been his fondness for smoking, drinking and gossiping with the local inhabitants, which by modern standards does not seem reprehensible; but it is indisputable that he made things difficult for the Cape Government by collecting complaints from the Dutch settlers and forwarding them privately direct to Europe. Hitherto such complaints, dispatched through official channels, had somehow got lost in the post. Yet he cannot have been altogether idle, and the two volumes he wrote about

Peter Kolben (Kolbe or Colbe), the first official Cape Astronomer.
– *Courtesy of Africana Museum, Johannesburg*

Cape natural history created quite a stir when they appeared, translated by Medley, in England in 1731.

Kolbe's observatory was situated in the Buren bastion of the Castle, which certainly explains some of the discrepancies in the latest of the three drawings purporting to show Father Tachard's observatory. No doubt he carried out useful work, but after five years experience of him the Governor of the Cape of Good Hope inscribed the following Resolution:

"The Astronomer, Pieter Colbe, who arrived here in the year 1705, in the ship called the *Unie* from Holland, and who has for a considerable time been idling about without prosecuting his astronomical observations or rendering any burgher-service . . . it was thought advisable to demand from him whether he intends to remain here much longer, in which case he will be considered as a burgher, and thus become liable to taxes and burgher-duty; otherwise we shall give him his discharge so that he may return to Europe."

The thought of going back to Europe and leaving the pleasures of the Cape behind him does not seem to have appealed to Kolbe, and he obviously chose to stay, as we next hear of him in Stellenbosch, occupying the post of secretary to the Court of the Landrost of the Heemraad. Here, presumably, he had less time for smoking, drinking and gossiping, since he left behind him an enormous mass of records in his own handwriting.

These two early astronomers of South Africa are rather shadowy figures. Only by studying the effects which they produced on other people – their contemporaries – can any

The Abbé Nicolas Louis de la Caille, 1713–1762.
– Courtesy of South African Library

assessment of their characters be even attempted. Fascinating though this may be, it seldom produces a reliable picture of the true flesh-and-blood person. The first southern astronomer about whom we know enough to be sure of recognizing him if we were to pass him in the street was Nicolas Louis de la Caille, who was born in 1713 at Rumigny in the Ardennes, not far from the Cathedral city of Rheims.

The Abbé de la Caille, as he was later to become, laid the foundations of southern-hemisphere astronomy. He was the predecessor of all the later pioneers such as Maclear, Herschel and Gill. It has sometimes been suggested that La Caille's achievements were largely vitiated by the lack of adequate instrumentation for his basic work; but a scientist of the stature of Sir Thomas Maclear did not think so, and neither did Sir John Herschel, whose survey of the southern sky is now generally held by astronomers to have been the true starting-point of modern scientific work in this field. Had La Caille been able to use modern-type instruments as well, his work would have advanced our knowledge of southern-hemisphere astronomy by almost a hundred years. As it was, he showed later generations how to set about an intensive observing programme, and his diligence in pursuing it has become legendary. It is not from pure sentiment that a piece of charcoal from one of his signal fires is preserved at the Cape Observatory to this day. Neither is it out of mere curiosity that the University of the Witwatersrand has acquired, by purchase, one of his notebooks – a treasure that it could ill afford. A picture of a page of the notebook is shown here.

Because of his importance, and because La Caille was such a fascinating character, let us examine the man and his achievements as closely as space considerations will allow.

De la Caille (often nowadays spelled "Lacaille") received the formal classical education of his time, and the religious title of Abbé was conferred upon him when he had completed his studies. No record can be traced, however, of any period during which he made the office of priest his sole concern. Indeed, the first extensive mention of him relates to the years when he worked with the brilliant Cassinis at the Paris Observatory.

The first Cassini, Giovanni – founder of the scientific dynasty – was the discoverer of four of the satellites of Saturn as well as the famous division in that planet's ring-system. He had arrived in the French capital from the University of Bologna in 1669 to take charge of the observatory which had recently been set up there on the express orders of the King. On Giovanni's death in 1712 (the year of La Caille's birth) his son, Jacques Cassini, succeeded his father as Director of the Observatory. It was through this distinguished scientist – he moved in the circle of Newton, Halley and Flamsteed, and was elected a Fellow of the Royal Society as well as receiving the highest scientific honours from his native country – that La Caille was given his first really responsible post.

His allotted task was to survey the French coast from Nantes to Bayonne. His success led, in 1739, to his being selected to remeasure the French arc of meridian. This operation, important in determining the size and shape of the Earth, was also completed to the full satisfaction of his sponsor. Yet La Caille's friendship with Jacques' son Caesar Cassini, who was a year younger and who was also engaged in important scientific pursuits, militated against, rather than for, La Caille's preferment. Jacques had no intention of laying himself open to the charge of nepotism. Luckily, his young protégé's total dedication and exceptional brilliance were obvious to all.

On completion of his second assignment La Caille was admitted to the French Academy

13 Sept 28952	Syrius	98 33 11,6					
	α Aquilæ	294 40 33,1	−10,5	−1 33,0	−0,6	+4,5	294 38 52,5
	β ♒	319 37 41,4	−16,0	−1 33,0	+0,2	−2,6	319 35 50,0
	α ♒	328 16 55,4	−16,9	−1 33,0	+0,0	−0,5	328 15 5,0
	γ ♒	332 12 57,9	−17,5	−1 33,0	−0,0	−0,7	332 11 6,7
	Phomalhaut	340 58 44,9	−21,2	−1 33,0	−0,7	−6,7	340 56 49,3
	γ Toucani	345 42 12,9	−36,1	−1 33,0	−3,2	−14,3	345 39 46,3
	β Hydri	3 5 18,6	−1 32,1	−1 33,0	−18,6	+9,1	3 2 4,1
	Achernar	22 7 8,6	−30,0	−1 33,0	[illegible],0	+20,9	22 5 17,5
	β Phœnicis	13 44 52,1	−25,2	−1 33,0	−5,3	+9,0	13 42 57,6
	γ Phœnicis	20 13 38,6	−24,7	−1 32,9	−6,4	+14,3	20 11 48,9
	δ Phœnicis	19 23 46,1	−22,6	−1 33,0	−5,2	+11,2	19 21 56,5
	Canobus	94 36 35,6	+8,2	−1 33,0	−7,7	+44,4	94 35 47,5
7 Octob.	Syrius	98 33 23,2					
	β Hydri	3 5 42,5	−1 32,0	−1 35,7	−19,8	+9,5	3 2 24,5
	α Phœnicis	3 29 51,5	−24,9	−1 35,7	−3,6	+2,1	3 27 49,4
	Algenib	0 7 48,0	−18,7	−1 35,7	+0,9	+0,0	0 5 54,5
	α Hydri	27 44 50,2	−41,4	−1 35,8	−11,5	+32,1	27 42 52,9
	* Eridani	45 23 30,5	−18,4	−1 35,7	−4,0	+14,6	45 21 47,0
	θ Eridani	42 13 19,5	−22,2	−1 35,7	−5,9	+20,9	42 11 36,6
	Canobus	94 36 58,7	−11,3	−1 35,7	−7,5	+46,1	94 35 50,3
8 Octob	Syrius	98 33 23,6					
	λ ♒	339 55 15,9	−15,0	−1 35,9	−0,3	−1,9	339 53 22,8
	ε Toucani	1 45 17,9	−44,2	−1 35,9	−7,9	+2,5	1 42 52,4
	Inf	9 59 53,1	−1 15,5	−1 35,9	−18,6	+25,2	9 58 8,3
	β Ceti	7 47 17,4	−19,4	−1 35,9	−1,6	+1,7	7 45 22,8
	β Toucani	5 1 43,6	−41,1	−1 35,9	−8,7	+6,5	4 59 24,4
5 nov.	Syrius	98 33 35,8					
	Achernar	22 7 21,8	−33,6	−1 39,2	−9,1	+22,7	22 5 22,6
		8 2 24,8	−28,2	−1 39,2	−7,3	+8,2	8 0 18,6
		39 27 28,3	−1 50,0	−1 39,2	−35,4	+133,3	39 25 37,0
		35 25 53,8	−1 18,5	−1 39,2	−16,2	+79,2	35 25 4,4
	γ Hydri	57 52 14,6	−30,6	−1 39,2	−10,4	+45,0	57 50 39,4
	α Reticuli	62 45 08,8	−40,4	−1 39,2	−13,7	+65,0	62 43 40,7
	* Eridani	62 8 20,8	−22,7	−1 39,2	−5,1	+23,8	62 6 37,6
	β Leporis	79 24 47,1	−17,1	−1 39,2	−2,6	+13,9	79 23 2,1
26 Octob	Syrius	98 33 31,6					
	α Dorad	67 10 22,9	−28,2	−1 38,1	−10,4	+49,0	67 8 55,2
	* Eridani	66 42 58,6	−17,1	−1 38,1	−2,6	+12,3	66 41 13,1
	υ Eridani seq	66 29 10,1	−18,2	−1 38,1	−4,2	+20,1	66 27 29,7
16 Octob	/	354 11 32,7					

A page of La Caille's notebook.

– Courtesy of the University of the Witwatersrand, Johannesburg

Dr Patrick Moore reading La Caille's notebook.
– *Courtesy of the University of the Witwatersrand, Johannesburg*

of Sciences, and was appointed to the Chair of Mathematics at Mazarin College, where a small observatory was established for him to use. The College Mazarin is now the Institute of France, and the name of Nicolas Louis de la Caille is of proud record in its archives.

It was with this background, and with several excellent works on mathematics and astronomy to his credit, that La Caille was singled out. And it was precisely because of his proven competence and wide knowledge that when he later made an error in one of his major geodetic measurements, men of the calibre of Maclear and Herschel went to enormous lengths to discover how the calculations of such an expert could possibly have gone wrong. Establish the cause of La Caille's mistake, they reasoned, and an important scientific principle could be unearthed. They were right.

In the middle years of the eighteenth century, lack of an astronomical station in the southern hemisphere was greatly hampering scientists in their basic work. First, a southern arc of the meridian was needed to complement the measurements already made in the northern hemisphere, and thus to determine the exact shape of the Earth. That the Earth must be oblate – that is to say, flattened at the poles – was generally suspected. Results from the north indicated as much, and if Newton's theory of gravitation were correct the southern half of the Earth should be the same shape as the northern half. But nobody could be sure; it had to be scientifically proved.

Then there was the urgent need for a reliable catalogue of the southern stars, to aid mariners in navigating their ships. The existing catalogues were, to put it kindly, sketchy as well as incomplete. Also, by working in conjunction with a second observatory in the north at approximately the same longitude, the exact longitude of the Cape itself could be established. Magnetic variation studies were also on the programme, as well as the measurement of the distances of the Moon and the nearer planets.

To its everlasting credit, the French Academy of Sciences was the first to take active steps. It had already sponsored such work in France, Lapland and Peru, and to this fact may be attributed the exhaustive nature of its preliminary preparations and its admirable definition of what such an expedition's aims should be. As it already had favourable experience of the work of one of its own members – the Abbé de La Caille – it is not surprising that the choice of a man to undertake the mission should have fallen upon him. The La Caille of that date – it was 1750, and he was 37 years of age – is shown in a contemporary picture as a man with regular features and a sensitive mouth. The eyes are kindly and humorous, their gaze direct. His attire is neat and sober, as befits his church rank, which is further stressed by his full-bottomed wig and clerical cravat. His attitude gives no hint of his almost incredible intellectual energy – but it was this energy which was the wonder of all who knew him, and which was to be marvelled at by famous scientists of later generations.

There is an old music-hall performers' adage which states: "Tell 'em what you're going to do, tell 'em what you're doing, and tell 'em what you done!" In more sober terms, this is what the French Academy of Sciences proceeded to do. The aims of the mission were published in detail long before La Caille set foot on the *Le Glorieux*, the ship which was to take him to the Cape; exhaustive reports were sent by him to his brother academicians during his two-year stay; and the speed with which his results were published on his return would have made those same music-hall artists swear that he had something up his sleeve.

The "expedition", in the event, consisted of La Caille himself.

A formal application was made by the Academy to the Dutch authorities in Holland, requesting permission for their distinguished member to proceed to the Cape in order to carry out his astronomical observations there. The States-General, after deliberation and consultation with the Dutch United East India Company, granted the required permission, and an entry in its journal dated 25 November 1750 reads:

"After deliberation on the subject, it was thought proper and resolved to permit M. De La Caille, as he is permitted by these presents, to proceed to the Cape of Good Hope for the purpose set forth in the said memorial, extract of this resolution of their High Mightinesses to be sent to the said Representatives in the Chamber of XVII at Amsterdam aforesaid, in order that the same may be communicated to the Governor of the Cape of Good Hope, and cause them to take such precautions as to prevent any improper use being made of this permission."

The wording of the final clause seems to indicate that Peter Kolbe had not been forgotten!

La Caille must have considered the granting of this permission to be a foregone conclusion, because even before it had been received he had embarked on the *Le Glorieux*, a vessel of 14 guns and 115 men, under the command of Captain Dapres. The voyage to Cape Town was made by way of Rio de Janeiro, from which port *Le Glorieux* sailed in February, anchoring in Table Bay on 19 April 1751.

Governor Ryk Tulbagh recorded the arrival in his official journal:

"Monday, 19 April 1751: Arrived the French ship *Le Glorieux*, carrying 14 guns and 115 men, which left Port L'Orient on 21st last year, under the command of Captain Dapres of Mannevillette, bound to the Island Mauritius, having on board Mr De La Caille, Member of the Royal Academy of Sciences at Paris, who will remain here for some time, with a view of making Astronomical Observations . . ."

La Caille lost no time in presenting his credentials, for he and Captain Dapres landed at ten o'clock on the morning of the 20th and paid a ceremonial visit to Governor Tulbagh, to whom La Caille handed the following letter from the Prince of Orange:

Sir, – Mr De La Caille, Member of the Royal Academy of Sciences at Paris, who intends to proceed to and remain at, the Cape of Good Hope for some time, with the view of pursuing his Astronomical Observations, having applied to their High Mightinesses, – and we also being disposed to comply with the request made to us in that respect by this gentleman, do therefore hereby recommend him to you, trusting that you will render him such assistance as he may be standing in need of during his temporary residence there.

Commanding you to God's holy protection and keeping,
We remain, Sir,
Your well-disposed friend,
Prince D'Orange and Nassau.
By order of His Highness,
In the absence of the Private Secretary,
J. D. Horst.

Op het Loo,
17 October 1750.

Governor Tulbagh was not to be outdone in the compilation of flowery phrases. The very next day he made this reply:

Most Illustrious Prince and Sir, –

Mr De La Caille, who arrived here on the 19th of this present month of April, having handed over to me your Highness's letter, dated Op het Loo, the 17th October last, I have, in dutiful compliance with the desire therein expressed, not only permitted that gentleman to remain at this place for some time for the purpose of proceeding with his astronomical observations, but I shall also not fail to render him such assistance as may be in my power towards bringing that work to a desirable conclusion; of which, I humbly trust, Your Highness is fully convinced, and also that I shall ever consider it as the greatest happiness

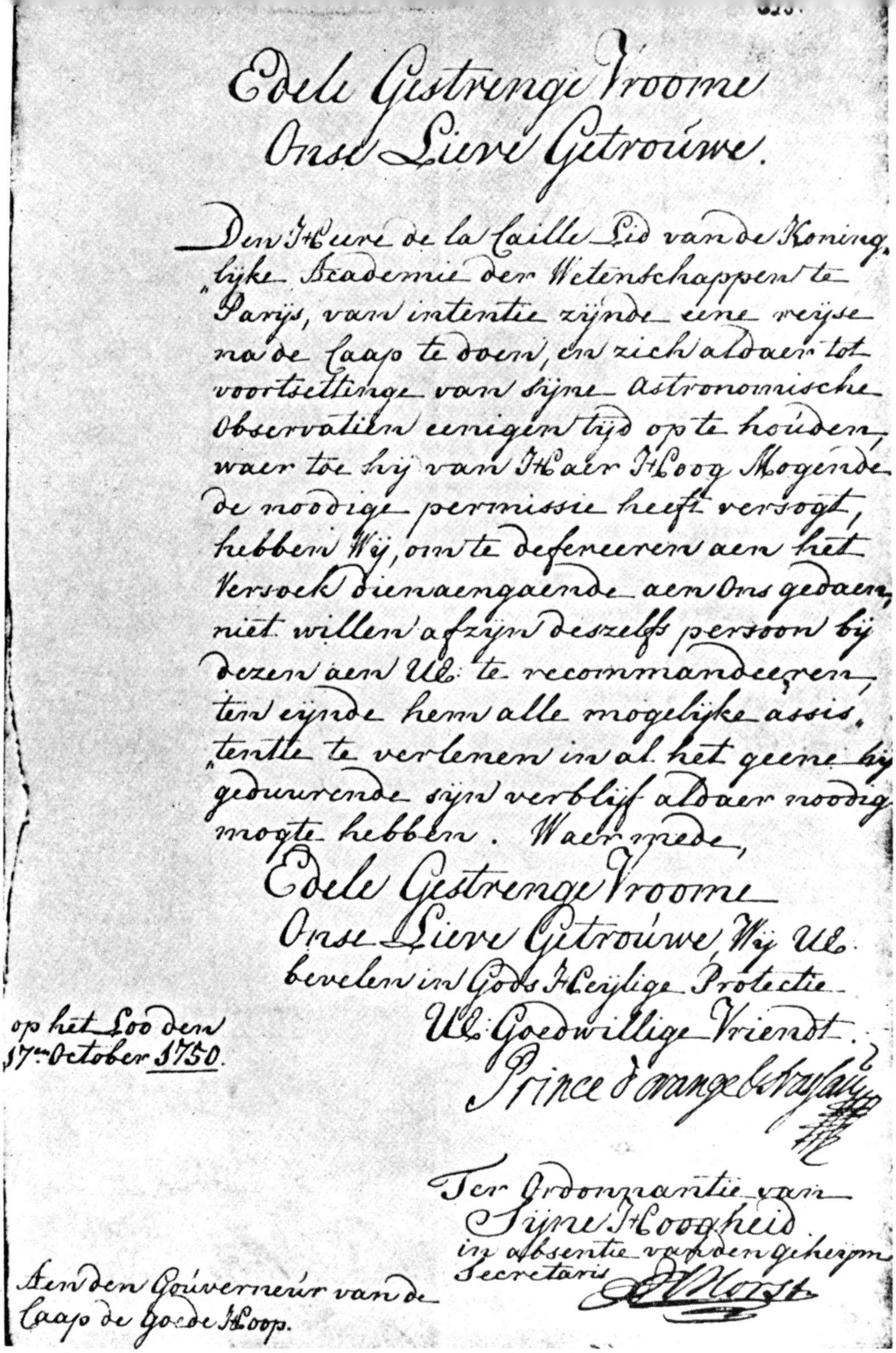

Edele Gestrenge Vroome
Onse Lieve Getrouwe.

Den Heere de la Caille Lid van de Koninglijke Academie der Wetenschappen te Parijs, van intentie zijnde eene reijse na de Caap te doen, en zich aldaer tot voortsettinge van sijne astronomische Observatien eenigen tijd op te houden, waer toe hij van Haer Hoog Mogende de noodige permissie heeft versogt, hebben Wij, om te defereeren aen het Versoek dienaengaende aen Ons gedaen, niet willen afzijn deszelfs persoon bij dezen aen UE: te recommandeeren, ten eijnde hem alle mogelijke assistentie te verlenen in al het geene hij gedurende sijn verblijf aldaer noodig mogte hebben. Waer mede,

Edele Gestrenge Vroome
Onse Lieve Getrouwe, Wij UE: bevelen in Gods Heijlige Protectie
UE: Goedwillige Vriendt.
Prince d'Orange et Nassau

op het Loo den 17en October 1750

Ter Ordonnantie van
Sijne Hoogheid
in absentie van den geheijmen Secretaris
[illegible]

Aen den Gouverneur van de Caap de Goede Hoop.

Letter from the Prince of Orange and Nassau recommending La Caille to the Governor of the Cape of Good Hope.

– Courtesy of the Cape Archives

Old houses in Strand Street, Cape Town (now demolished). La Caille resided in the house on the right.

– Courtesy of South African Library

in having an opportunity of subscribing myself, with sentiment of profound respect and esteem,

Most illustrious Prince and Sir,

Your Highness's most faithful, most obedient and most humble servant,

R. Tulbagh.

In the Castle of Good Hope,
the 21st April, 1751.

Memorial plaque, designed by Sir Herbert Baker, on the site of La Caille's Cape Town station, showing his triangulation for measuring an arc of meridian, his observing instruments, etc.

– Courtesy of South African Library

It would have been maddening for a letter in these terms to be lost in the post in the manner of some of the Burgher's complaints – so the Governor sent it off in duplicate, one copy by the ship *Overschré* and the other by the ship *Peregtigheijd.* Both copies arrived.

Sadly, the house in which La Caille stayed in Cape Town has now been demolished. Its site is marked by a plaque. The residence of Jan Laurens Bestbier, it was situated in Strand Street – "the second house counting westward from the Heerengragth" (now Adderley Street) "on the side next the Bay". According to a later observer it was "of the description of a good London house". Sir David Baird, Lord Glenelg and Prince Frederick of Orange had lodged in it, and Captain Cook was a later guest.

La Caille erected a small observatory in the courtyard behind the house, at the end nearest the sea. The telescope he used was 28 inches long and half an inch in aparture. Modern amateurs would scorn an instrument of similar size. Yet with it La Caille accomplished a prodigious amount of work, in volume equal to the research programme of an important observatory of the present day.

He remained at the Cape for less than two years, and in that time he charted the positions of almost 10 000 stars, grading them according to their brightness. He added a number of new constellations and gave them names – names which, sometimes in shortened form, have been retained by modern astronomers. Instead of the classical nomenclature of mythology, he used scientific instruments as the inspiration for his new titles. Thus we have:

Apparatus Sculptoris (the Sculptor's Tools)
Fornax Chemica (the Chemical Furnace)
Horologium (the Clock)
Reticulus Rhomboidalis (the Rhomboidal Net)
Cæla Sculptoris (the Sculptor's Chisel)
Equuleus Pictoris (the Painter's Easel)
Pyxis Nautica (the Mariner's Compass)
Antlia Pneumatica (the Air-Pump)
Octans (the Octant)
Circinus (the Compasses)
Norma, *alias* Quadrans Euclidus (Euclid's Square)
Telescopium (the Telescope)
Microscopium (the Microscope)
Mons Mensæ (Table Mountain)

His chart of the southern constellations is shown here.

La Caille did not live long enough to reduce all his observations. This work was later undertaken by British astronomers, and publication of *A Catalogue of 9 766 Stars in the Southern Hemisphere . . . from the Observations of the Abbé De La Caille* was financed by the British Association for the Advancement of Science in London, in 1847.

Not only did La Caille chart the southern stars, but in 1755 he also dispatched to the French Academy a catalogue of 42 nebulæ which he had observed during his stay at the Cape. Moreover, he kept a daily record of weather and tides while living in Strand Street, and he made regular magnetic observations.

Important as all these observations were, his basic work in the field of surveying was

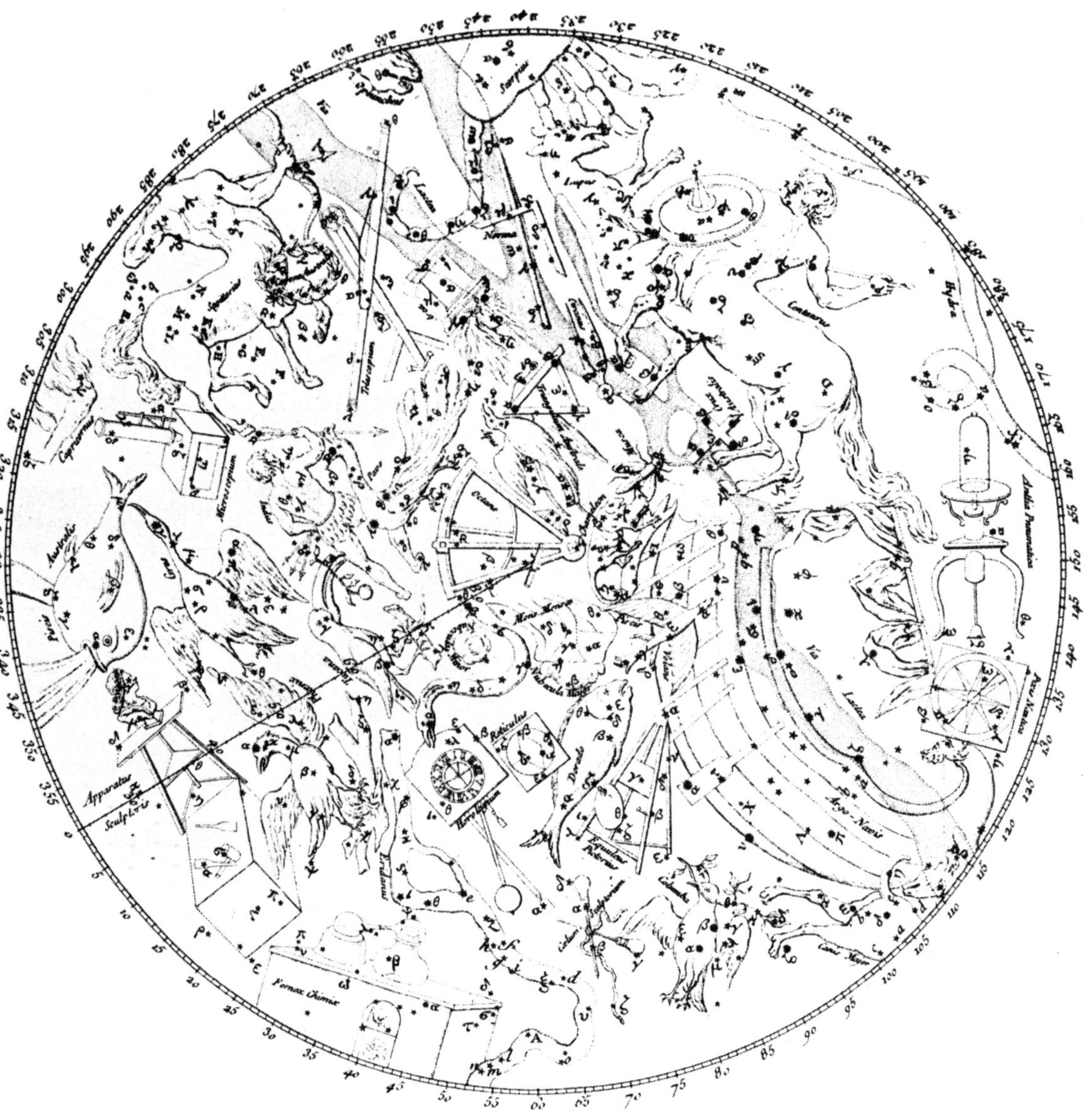

La Caille's Constellations.

– Courtesy of South African Library

equally vital. In their published statement of the aims of the expedition the French Academy had claimed that he would accurately determine the position of the Cape, and also measure an arc of the meridian in the southern hemisphere, in order to give a final value for the shape of the Earth. La Caille managed this successfully by using observations of the satellites of Jupiter to fix the longitude of the Cape. Measuring the arc involved much field work; and even while he was busy on his star catalogue he found time to pay several visits to the country north of Cape Town to find suitable sites for his proposed work. A description of how he set about it is of interest even today.

His preliminary excursions had suggested that it would be comparatively simple to measure an arc of the meridian of approximately $1\frac{1}{4}$ degrees in length. An arc of the meridian simply means a single degree of longitude along the Earth's surface; the measurement can be found only by astronomical observations. In the northern hemisphere, measurements had shown that a degree of longitude tends to lengthen as one approaches the North Pole. Therefore, in the northern hemisphere at least, the Earth cannot be a true sphere; it is a sphere which is slightly flattened at the North Pole.

For verifying many of his Cape observations, La Caille had to know the exact distance between the latitude of his Strand Street observatory and the latitude of the Paris Observatory. To measure this with sufficient accuracy, he had to know the true shape of the Earth.

In August 1752 he made a preliminary field trip to select a suitable site for the northern point of his arc. The southern point, he had already decided, would be his observatory in Strand Street. This was later to prove an unfortunate choice.

The northern point he selected was Klipfontein, a farm in the mountains north of Piquetberg. His fundamental baseline, which was eight miles long, was measured on the Darling Flats, using wooden rods. On this base, by observations of angles, he built up a series of triangles connecting his Strand Street observatory to Capocberg, not far from Darling; Riebeek Casteel, and Klipfontein. The true distance between Klipfontein and Strand Street (the north and south stations) could now be established by calculation. La Caille then made astronomical observations to fix the latitude difference between the two stations. La Caille's own survey of his arc is shown on his map, reproduced here.

Of course, it did not all go smoothly – things never do, on expeditions of this sort, which is why they are so fascinating to discuss afterwards. True to his promise to the Prince, Governor Tulbagh loaned La Caille some men, under the command of Captain E. B. Muller, an artillery officer who was also the engineer at the Cape Fortress. Also mentioned in the reports is a man named Poitevin, held by some to have come from France with La Caille – though we have found no proof of this in the documents which we have examined.

The party was held up for eleven days and nights on Riebeek's Casteel by bad weather, and the beacon fires on Piquetberg which were to be used as fixed points were not kept constantly alight, as they should have been. Not surprisingly, on 11 October 1752 Captain Muller wrote to the Governor requesting a few more days' leave. One can well imagine the conscientious La Caille's frustration at the delay, but the support he received from Muller must have been a great solace to him. It has been said of the worthy Abbé that he won the love and friendship of all who knew him. Certainly this seems to be borne out by the many records we have of official and private persons rallying spontaneously to his help.

The result of La Caille's measurement, however, provided a real shock. From it – and he checked his figures with his customary care – the southern half of the Earth was found to be of different shape from the northern half. It was a prolate, or slightly lengthened spheroid instead of an oblate, or slightly flattened one! As we would express it today,

La Caille's sundial in the courtyard of number 7, Strand Street. "A brass plate perforated with a small hole, and fixed horizontally on a vertical wall, with a black line traced immediately below it, for the obvious purpose of determining the sun's passage over the meridian, still stands."
– *Courtesy of South African Library*

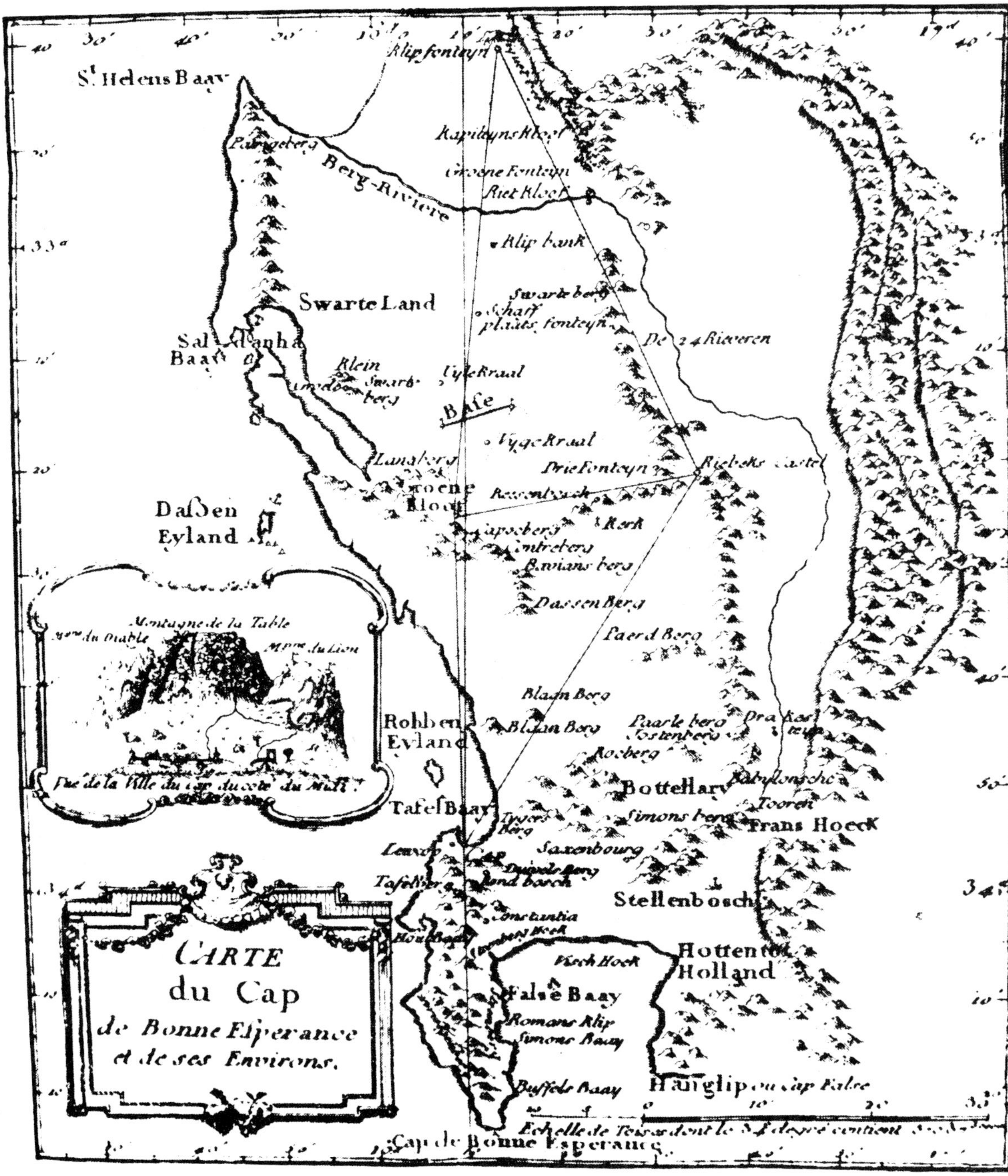

La Caille's map of his triangulation for measuring an arc of meridian. From his *Journal Historique*, 1763.

– *Courtesy of South African Library*

ASTRONOMIÆ

FUNDAMENTA

NOVISSIMIS

SOLIS ET STELLARUM

OBSERVATIONIBUS

STABILITA

LUTETIÆ IN COLLEGIO MAZARINÆO

ET IN AFRICA AD CAPUT BONÆ SPEI

PERACTIS

A Nicolao-Ludovico DE LA CAILLE, in almâ Studiorum Univerſitate Pariſienſi Matheſeon Profeſſore, Regiæ Scientiarum Academiæ Aſtronomo, & earum quæ Petropoli, Berolini, Holmiæ & Bononiæ florent, Academiarum Socio.

PARISIIS;

E Typographiâ J. J. Stephani COLLOMBAT, Typographi ordinarii Regis.

M. DCC. LVII.

Title page of La Caille's *Astronomæ Fundamenta.*

– *Courtesy of Africana Museum, Johannesburg*

La Caille's single degree of longitude, or arc of the southern meridian, measured 111.17 km. instead of 111.30 km., its anticipated value – a discrepancy of 0.13 km.

Of course, from the lofty pinnacle of hindsight, we know that La Caille's measurement was wrong – and, moreover, we know why. But it took a surveyor of equal stature, a staunch admirer of La Caille, to make the correction, as will be described later.

His measurement of the arc of the meridian rounded off La Caille's work at the Cape, and on Thursday 8 March, 1753, he sailed for Mauritius aboard the French ship *Le Puisieux*, none too elated, it is said, by the fact that he was not going straight home to France. When he finally returned to Europe and the College Mazarin at Paris he deduced a number of important results from his southern observations. His contributions to science were among the most valuable to be made during the whole of the eighteenth century.

He died at the age of 48, burned out by the intensity of his intellectual fire.

In the years immediately following La Caille's visit, little of astronomical note seems to have happened anywhere in South Africa. A few illustrious names crop up in the records – for example, those of George Mason and Jeremiah Dixon, two English engineers who were later to give their names to the Mason-Dixon line, popularly supposed to have been a line dividing the slave-holding from the non-slave-holding states of America. But this work lay in the future. Mason and Dixon went to Cape Town in 1761 specifically to observe the transit of Venus. This they did most successfully from a temporary observatory in Concordia Gardens, a public pleasure garden which, on a smaller scale, was modelled on London's Vauxhall and Cremorne. It was sited in Cape Town, behind the place where St Mary's Cathedral now stands.

Wales and Bayly, the astronomers of Captain Cook's second voyage with the vessels *Resolution* and *Adventure*, which sailed from Plymouth on 13 July 1772 to find out how far the land of Antarctica stretched northward, made observations from the same site.

Captain George Everest, the great English surveyor after whom the world's highest mountain was later to be named, called at the Cape on his way to India in 1820. He tried to identify the site of La Caille's Strand Street observatory, but was unable to do so with any certainty – although, later, he succeeded.

Then, in that same year – 1820 – the Board of Longitude in England pressed successfully for the setting-up of a permanent observatory in South Africa. The advancing character of every branch of astronomy demanded better and more frequent observations. Not unexpectedly – it was to happen again in Johannesburg – uninformed local opinion, ignorant of the true issues involved, opposed the scheme, putting forward, in this instance, the need for a new breakwater as being much more urgent. Luckily, an Order in Council clinched the matter, and the first Cape Astronomer was appointed: the Rev Fearon Fallows.

3. THE CAPE: YEARS OF PROGRESS

FEARON FALLOWS is an important figure in the story of South African astronomy. As it is to him that the Royal Cape Observatory owes the selection of the original site – a selection which was later to arouse much bitter criticism – it is interesting to find out what kind of a man he was, and why that controversial site was chosen.

Fallows was born at Cockermouth, in Cumberland, on 4 July 1789. He came of a family of hand-loom weavers, and as a boy he was apprenticed to that trade. His father was of scholarly disposition, and, appreciating the boy's quick intelligence and aptitude for learning, made sure that his son had the best education that could be obtained for him.Both father and son were interested in mathematics, and even before formal tuition on the youth's behalf was secured they made considerable progress in that subject.

On the recommendation of the Rev. H. A. Hervey, vicar of the neighbouring parish of Bridekirk, the young man – he was now 19 years of age – was appointed assistant to the headmaster of Plumbland School. So well did he carry out his duties, and so great was his progress in his own studies, that a patron readily financed his further education at St John's College, Cambridgé. He graduated in 1813 as Third Wrangler; in the same year Sir John Herschel, whom we will shortly meet, was First Wrangler. ("Wrangler" is the name given in the University of Cambridge to those who have attained the first class in the public mathematical honours examination. The word itself is derived from the public disputations in which candidates for degrees were, in former times, required to exhibit their powers.) Fallows obtained his M.A. degree in 1816, and after two years as a lecturer in mathematics at Corpus Christi College he was elected a Fellow of St John's. It was also during this period that he was ordained in the Church of England.

He had no experience of practical astronomy – indeed, it is said that he had never looked through a telescope – but on his appointment to the Cape post he embarked on what is nowadays known as a crash-course, visiting observatories and similar installations, and making himself familiar with the use of astronomical instruments. On 1 January 1821 he married Mary Anne Hervey, eldest daughter of his first benefactor, the Rev. H. A. Hervey. The couple arrived in Cape Town on 12 August of that same year.

Fallows' first task was to select a suitable site for the observatory, and there were various hazards to be borne in mind. The worst of them was sand – sharp, cutting, silicious sand, blown by the winds and whirlwinds of the many storms which affect the Cape. Therefore, Fallows had to find a place which was as free as possible from the effects of abrasive sand. Since one of his main objectives was to give signals and weather reports to ships in Table Bay, the observatory had to be easily visible from the decks of such vessels. The nature of the surrounding horizon – clear or obstructed – was another important factor.

Never a man to rush a task of this kind, Fallows installed his wife and himself at *Zorg-en-lust* in the Gardens until he had made his choice. He erected a temporary observatory, and began the astronomical work with which he had been entrusted. This was to make observations to improve the theory of refraction, reporting the results in the form adopted by Greenwich; also, to compile a new catalogue of the southern stars.

After much deliberation, he finally selected a site on rising ground four miles east of Cape Town, on a low tract of land which connects the mountainous peninsula on which

Cape Town itself stands, together with Table Mountain, to the main continent. The observatory was duly built, on a greenstone foundation with a firm upper soil of clay. It stood like an island amid all the evils that Fallows had been at pains to avoid. The cloud of red dust, seen daily as it rolled along the high road on its way into Cape Town, left the observatory untouched. The mighty clouds of white sand, blasting their way across the plains of the flats and destroying all the vegetation in their paths toward Table Bay, approached from the other direction – and they too left the observatory and its grounds unscathed. It was true that the land at the foot of the prominence was marshy, lying as it did between two rivers; moreover there were hippopotami in the rivers, and the original name of the hill, Slangkop or Snake Hill, indicated another form of life which could well be encountered. Leopards were also in evidence. Some of the later astronomers said hard things about Fallows, and one of them – his immediate successor – nicknamed the site the "Dismal Swamp". And yet it was the best that could be done, and the horizon was almost unobstructed, except for part of the west.

The building, based on a design by J. Rennie, was completed in June 1827. The instruments, which included a transit circle and a mural circle, were installed by the end of 1828 – seven years after Fallows and his bride had arrived. The couple took up residence in that part of the building which is now called Fallows House, and its large drawing-room was used as the first Anglican chapel in South Africa.

Generally speaking the assistants were of poor calibre, and Fallows often had to enlist the help of his wife (in 1830 she discovered a comet in the constellation of Octans). Conditions were never easy. In September 1830 the entire staff went down with scarlet fever, and Fallows' chief helper, Captain William Ronald, was sent home to Britain. Weakened by illness and by his labours, the first Cape Astronomer died on 25 July 1831, and, at his own request, was buried in the Observatory grounds; he was only 42 years old. Mary Anne went home to England, taking with her copies of her husband's observations. The reductions of these observations were completed and prepared for the press by Sir George Airy, the Astronomer Royal.

Fallows' successor, Thomas Henderson, detested the place from the start. He it was who named it the Dismal Swamp. He stayed for thirteen months, and ever afterwards maintained that this was a year too long. Yet it is a mark of his exceptional qualities that during his brief tenure of office he made five or six thousand observations of the places of southern stars, observed Encke's and Biela's Comets and the transit of Mercury of 5 May 1812; he studied stellar occultations and the eclipses of Jupiter's satellites, as well as making special parallax-observations of Mars and the Moon.

However, Henderson's most famous work was in connection with the distances of the stars – or, rather, of one star in particular: Alpha Centauri, the brighter of the two Pointers to the Southern Cross. His investigations were of fundamental importance, and it was only by sheer bad luck that he missed the honour of making a great discovery.

It had long been known that the stars are extremely remote. Men such as James Bradley, the third Astronomer Royal of England, had tried to measure their distances by what is termed the method of parallax – the slight apparent shifting of a comparatively nearby star over a period of six months, due to the real movement of the Earth from one side of its orbit round the Sun to the other. What had to be done was to measure the extremely small angular displacement of the target star against the background of stars which were farther

Henderson's "Dismal Swamp" in 1911.

– *Courtesy of Africana Museum, Johannesburg*

away. Bradley had failed; so had no less a person than Sir William Herschel – though in each case the attempts resulted in a discovery of quite a different sort: in Bradley's case the aberration of light, in Herschel's the detection of binary or physically-associated double stars.

One of the most promising stars was Alpha Centauri, because it was a wide binary and because it had unusually large individual or proper motion. His measures were made at the Cape, but he returned home before he had reduced them, and he delayed for several years. In 1838 Friedrich Bessel, at Königsberg, announced that he had carried out parallax measurements on a much fainter star (61 Cygni, in the Swan) and gave its distance as approximately 11 light-years.* It was only then that Henderson gave his results

*A light-year is the distance travelled by a ray of light in one year; it is equal to rather less than 6 million million miles, or about 9 million million kilometres.

for Alpha Centauri, whose distance proved to be much less – a mere 4.3 light-years, or under 25 million million miles. Henderson's estimate was very accurate, and Alpha Centauri is now known to be the nearest star in the sky – excluding the Sun, and also excluding Proxima Centauri, which is a faint companion of Alpha Centauri and which lies about 1/10 of a light-year closer to us.

It cannot be denied that the honour of priority must go to Bessel, but it is equally true that Henderson's measurements were carried out first, so that the discovery was within his grasp. Not that there were any personal recriminations; the two astronomers were on the best possible terms, and later on they went together on a holiday trip to the Scottish Highlands.

By South Africans, Henderson is sometimes regarded with a certain reserve, because he made no secret of his dislike of the surroundings in which he was placed. It may help the critics to understand Henderson better if we state that had he been appointed to the Cape a hundred and fifty years later he would probably have been able to cope with the situation easily enough. He was suffering from incipient heart trouble, and doctors were scarce. He became depressed, and in 1833 he wrote to a fellow astronomer as follows:

> "I will tell you about my residence in the Dismal Swamp among slaves and savages – plenty of insidious, venomous snakes. What would you think if, on putting out your candle to step into bed, you were to find one lurking beside the bed?"

Henderson resigned his post in May 1833 without troubling to make more than a flimsy excuse, and lost no time in returning to his native Scotland. On 1 October 1834 he became the first Astronomer Royal for Scotland, and was also appointed to a professorship at Edinburgh University, a post which was combined with the directorship of the observatory on Calton Hill.

Born at Dundee in 1789, the youngest of five children of a respectable tradesman, Henderson was said to acquire scientific knowledge by intuition. His contemporaries described him as an upright, benevolent and enthusiastic person. His wife pre-deceased him in 1842, and when he himself died of heart disease two years later his daughter was left with only the proceeds from the sale of his library to support her. True, it was a superb one, and it fetched a fair price. True, also, that he knew his daughter would not be left destitute, because he seems to have had foreknowledge of his brother-in-law's benevolent intentions. Not long after Henderson's death, the girl's uncle left her a large fortune.

There is no known portrait of Henderson, which is a pity; one would dearly like to know more about his personal appearance. Historians of astronomy are always hoping that a picture of him will be found one day.

In 1834 the post of H.M. Astronomer at the Cape was allotted to Thomas Maclear. To this man South Africa owes a great deal, for it was he who put the Observatory on a really firm footing, as well as carrying out important work on his own account and paving the way for his brilliant successors.

In spite of a strangely improbably early career for such a position – he was a medical doctor by profession, and an astronomer only by inclination – he devoted himself wholeheartedly to his task. He was no Henderson, to flit in and out of the place like a firefly – though it must be added that he knew Henderson well, and liked him. When Maclear

Sir Thomas Maclear. ("A good Irish face").
– *Courtesy of South African Library*

arrived in Table Bay it was with the firm intention of staying there for as long as he could be of service. In the event, this proved to be for a period of 36 years.

Maclear was born at County Tyrone, in Ireland – he has been described as having a good Irish face – in 1794, and was originally destined to enter the Anglican Church. At an early stage he decided that this would not give him enough scope for his scientific interests, and he gladly agreed to the suggestion of two distinguished medical uncles, Sir George and Dr T. Magrath, that he should become a doctor. After qualifying, at Guy's and St Thomas's, he was appointed house surgeon at Bedford Infirmary, where he was a distinct success. Indeed, he appears to have been as good a doctor as he was later to become an astronomer and surveyor.

In 1825 he made a happy marriage to Mary Pearse, daughter of Theed Pearse, Clerk of the Peace for the county of Bedford. By now he was in partnership with one of his uncles at Biggleswade, and the practice prospered. Yet a complete change in his life lay close ahead.

While living at Bedford, Maclear had made the acquaintance of the Smyth family, whose head, Admiral Smyth, had served in the Royal Navy during the Napoleonic Wars. The old Admiral was very much of a character – to put it kindly, he was an eccentric.

When in command of the Anglo-Sicilian fleet at Messina he had made hydrographic surveys of the area without consulting his superiors, as technically he ought to have done. This passed without much comment from their Lordships of the Admiralty, who indeed may have thought it quite a good idea. Smyth married the daughter of a British merchant who lived in Naples, and naturally made various Italian friends. One of them was the astronomer Guiseppe Piazzi, who had acted as director of the Palermo Observatory in Sicily, and had achieved fame in 1801 when he had discovered (and named) the first asteroid or minor planet, Ceres.

It was from his meeting with Piazzi that the Admiral's interest in astronomy dated. On his retirement to Bedford he armed himself with the best instruments he could find – and they were very good indeed – and became a Fellow of the Royal Astronomical Society. He also published some atronomical books, one of which, the *Cycle of Celestial Objects*, is still regarded as something of a classic. To quote a contemporary reviewer: "The descriptions of the various objects are enlivened with a vast amount of general classic and antiquarian lore, introduced in the most genial spirit."

When the Admiral's next son arrived, the name chosen was Charles Piazzi Smyth. This was predictable; but less predictable was the fact that for a long period this same son would become a staunch assistant to Maclear at the Cape – and that eventually he would become the second Astronomer Royal for Scotland. Charles Piazzi Smyth was a strange mixture. He was an excellent practical astronomer, but he was also obsessed by "pyramidology", and he believed that mankind's destiny had been foretold for those able to read it by the arrangement of the chambers and other features of the Great Pyramid in Egypt. When analyzed, the arguments he put forward are shown to be weird in the extreme; nevertheless, they started a cult which has never really died out.

At any rate, Maclear and the Admiral struck up a friendship, and while Maclear was at Bedford he was given free, unrestricted use of the Admiral's astronomical equipment. He began observing, and his results were so good that the professionals soon took notice of him. He became a Fellow of the Royal Astronomical Society, and frequently attended London meetings, so that his circle of scientific friends increased. His Cape appointment followed; but his partner in the medical practice, his uncle, thought him mad to accept it. When Maclear asked for a relatively small loan so that he could pay his fare to South Africa, his uncle flatly refused to help in what he thought would be the ruin of his nephew's life.

Finally Maclear secured a passage on a Government ship for the sum of £36, which sounds little enough today even though when all the essential extras were tacked on it rose to around £220. The ship was the *Tam o' Shanter*, and the voyage took only five days less than three months. After the voyage, one of his five children died. On arrival he was met by Lieutenant Meadows, who had been Henderson's assistant and was now to be his. Meadows' greeting was not encouraging: "So, Sir, you have decided to accept this wretched appointment!" The date was 5 January, 1834.

Yet from these inauspicious beginnings, Maclear was to attain such stature that a South African town and district were to be named after him – and, of course, he was to receive a knighthood.

With the Maclears, in the rôle of valet, came Thomas Bowler, who was to prove an indifferent to poor servant but an excellent visual recorder of Cape buildings and scenes

Sir John Herschel.
– *Courtesy of Africana Museum, Johannesburg*

during the years to come. His artistic technique may have left something to be desired, but his charming, unselfconscious water-colours and line drawings are now valued items of Africana. In many cases they are the sole records of the places he depicted.

This seems to be the moment to introduce a man who was to become a staunch friend to all the Maclears, and a valuable partner to Thomas in many of his early scientific undertakings. This man is Sir John Herschel, one of the most interesting personalities in the history of astronomy – and certainly one of the most brilliant.

As the son of Sir William Herschel, who discovered the planet Uranus and compiled catalogues of northern-hemisphere double stars and nebulæ as well as providing the first reasonably accurate scheme of the shape of the Galaxy, John began life with every possible advantage. After graduating from Cambridge he worked with Sir James South in re-examining the nebulæ and clusters described in his father's lists, making 525 additions to the catalogue. His scientific achievements were so many and so varied that to mention them all would take a great deal of space, but at least a brief account of him must be given here.

At Cambridge he was Senior Wrangler (Fallows was Third) and First Smith's Prizeman. An early publication, at the age of 28, was *A Collection of Examples of the Application of the Calculus of Finite Differences.* In 1826 the Royal Society awarded him and South a gold medal apiece for their observations of double stars. His treatises on sound and on the theory of light appeared in the *Encyclopædia Metropolitana* in 1830 and 1831; his treatise on astronomy and his "preliminary discourse on the study of Natural Philosophy" in Lardner's *Cyclopædia.* Throughout his life he went on contributing valuable papers to scientific journals. He took an active interest in the development of the Royal Astronomical Society (originally the Astronomical Society of London) and served as its President. The range of his scientific interests was remarkable. For instance, his researches on the undulatory theory of light would alone have brought him fame. Moreover, he made important progress in the new science of photography, independently of Fox Talbot. He discovered – and described to the Royal Society on 14 March 1839 – the photographic importance of sensitised paper, and he was the first to use "hypo", or hyposulphite of soda, as a fixing agent. It has been said that he was the first to coin the word "photography", and it is quite certain that when you use the terms "positive" and "negative" as they relate to photography you are recalling the terms which Herschel invented.

Obviously, Herschel was amazingly versatile; a true Admirable Crichton of advanced scientific studies. The poet Campbell described him as a prodigy, and this was very true. Yet, with it all, he never lost his amiability, his ready friendliness, or his natural charm. Even his tendency to fuss seems to have enhanced, rather than diminished, the affection felt for him by his family and friends. On 3 March 1829 he married a Scots girl, Margaret Brodie Stewart, and their union was described as being "of unclouded happiness".

In 1831 King William IV created him a Knight of the Royal Hanoverian Guelphic Order. By 1833 his pre-eminence was beyond the necessity of being marked by honours; but the honours continued to pour in.

Having completed and extended his father's catalogue of the northern heavens, Herschel arrived at the Cape, in January 1834 (ten days after Maclear) with the set intention of making a similar catalogue for the southern hemisphere. He bought an estate at Feldhausen, six miles from Cape Town, erected his 20-ft. focus reflector and a 7-ft. refractor, and was ready for work before June.

It is often supposed that Herschel was in some way officially connected with the Royal Observatory. He was not; but his presence during the early years of Maclear's work there was invaluable. The two men became fast friends, a friendship founded on mutual respect; it is not too much to say that had it not been for Herschel, with his constant stream of helpful suggestions and his warm personality, the first four years of Maclear's work would have been much less pleasant than they actually were – and considerably less fruitful.

At this stage Herschel was world-famous, while Maclear was at the start of his astronomical career. It has been said that at the beginning of their association it was Herschel who provided the necessary impetus, while Maclear was still groping uncertainly to decide exactly where his most useful work might lie. Later, the rôles were somewhat reversed, as

Photographic reproduction of a camera lucida sketch by Herschel of his 20-ft telescope *in situ* at Feldhausen, 1834.

– *Courtesy of Africana Museum, Johannesburg*

Maclear's self-confidence grew. Whatever may be the truth of it, the two men remained firm friends to the end of their lives.

Herschel described his house and estate at Feldhausen as occupying "one of the most magnificent sites I ever saw". The site is breath-taking even today; and a visit to the Herschel Obelisk – about which more will be said later – is well worth while. The monument stands on the exact spot where the 20-ft. reflector was set up, just to the east of the Table Mountain massif. From this viewpoint Devil's Peak looms majestically in the distance, a circumstance which, for all its beauty, was to prove somewhat of a nuisance in Herschel's cometary observations.

The homestead itself was demolished as recently as 1958, but from contemporary pictures and descriptions an excellent idea of it may be conjured up. It was built in the Cape Dutch style with a thatched roof, a central gable, and a wide stoep with seats at either

A highly-finished camera lucida sketch by Sir John of Devil's Peak, as seen from the stoep of Feldhausen.

– Courtesy of South African Library

end. Entrance to the house was gained by a flight of broad steps, flanked by enormous oaks. The walls, built of cut river-stone and sun-baked brick, were two feet thick, providing warmth in winter and coolness in summer. Heavy stinkwood joists supported the floors, and yellowwood doors with handwrought iron hinges added to the general charm. The floor of the spacious entrance hall was of grey Batavian flagstones.

The outbuildings included servants' quarters, stables, a cowshed, and a circular dairy. A cottage in the grounds was pressed into service as Sir John's workshop, where he polished his mirrors, conducted chemical experiments, and spent hours in his studies of the colouration of flowers. Two hexagonal sentry-boxes were situated at strategic points. One of these exists today, the sole forlorn survivor of all the gracious buildings which Sir John and Lady Herschel knew.

Camera lucida sketch by Sir John of his daughters Caroline, Isabella, and Louisa in the north-east avenue of Feldhausen.

– Courtesy of South African Library

An impressive stone-pine avenue led eastward, and a cross-avenue of oaks stretched to the edge of the vineyard. The estate was of 99 morgen, a goodly size even in those days.

For reasons which remain something of a mystery – after all, his home was six thousand miles away – Herschel bought the estate in 1835 for the sum of £3 000. Laudable attempts have been made to estimate how much this figure would represent in modern currency, but in these days of rocketing land values the calculation depends on so many imponderable values that it is almost impossible to make. It is only necessary to add that a 10 lb. fish could be bought for 2d, while butcher's meat was $1\frac{1}{2}$ d per lb.! Fortunately, Herschel's parents had left him well off. When he returned to England, the Government offered to reimburse him for the money he had spent on his South Africa expedition. Herschel refused.

Initially, the Feldhausen ménage consisted of Sir John and Lady Herschel with their three young children, Caroline, Isabella and baby William, and John Stone, a mechanic who had come with the family from England. To these were added, on arrival: Somai the coachman, David the groom, Minto the houseboy, Dawes and Jack the labourers, Jephtha and his wife as cooks, Clara and Candassa as washgirls, Nancy the needlewoman, a herdsman, and several other servants whose duties seem to have been of a less specific nature. Three more children were born to the Herschel's during their stay – Margaret, Alexander and John. Eventually the Herschels had three sons and nine daughters, all of whom remembered their parents as being devoted and impartial.

Having introduced Sir John, let us return to Thomas Maclear.

Maclear had lost no time in putting the Observatory in order, and establishing a routine which soon taxed the capacities of his limited staff. He threw himself with enthusiasm into an observing programme, which meant a lag in the reductions; indeed, some of the observations had to wait for reduction by Maclear's successors, Stone and Gill. The ultimate value of his work was, of course, very great. He was fully conscious of the fact that a large volume of accurate observations was of prime importance at that stage, so that a delay in reducing them was by no means disastrous.

As well as his astronomical work, Maclear took part in land surveys as well as magnetic, meteorological and tidal observations. His most famous surveying work was the remeasurement of La Caille's meridian arc, in an attempt to find out if – or where – that great man had gone wrong. To help in this undertaking – in which Herschel gave invaluable advice throughout – the Astronomer Royal, Sir George Airy, sent out from Greenwich the zenith sector which had been used by James Bradley, discoverer of the aberration of light; Bradley had died 75 years before. A reproduction of the first page of Airy's letter of instructions to Maclear is shown here. For the benefit of those who find difficulty in deciphering Airy's handwriting, this is what he says:

"Draft of instructions to Mr Maclear, Astronomer at the Royal Observatory, Cape of Good Hope, in reference to the use of Bradley's Sector for the verification of La Caille's arc of meridian.

"I. The object of the proposed observations is to verify La Caille's observations, and to explain if possible the cause of the discordance of La Caille's result from the results of all similar observations. For the first of these purposes, it is indispensable that the observations be of unexceptional goodness, for which it is presumed that the admirable construc-

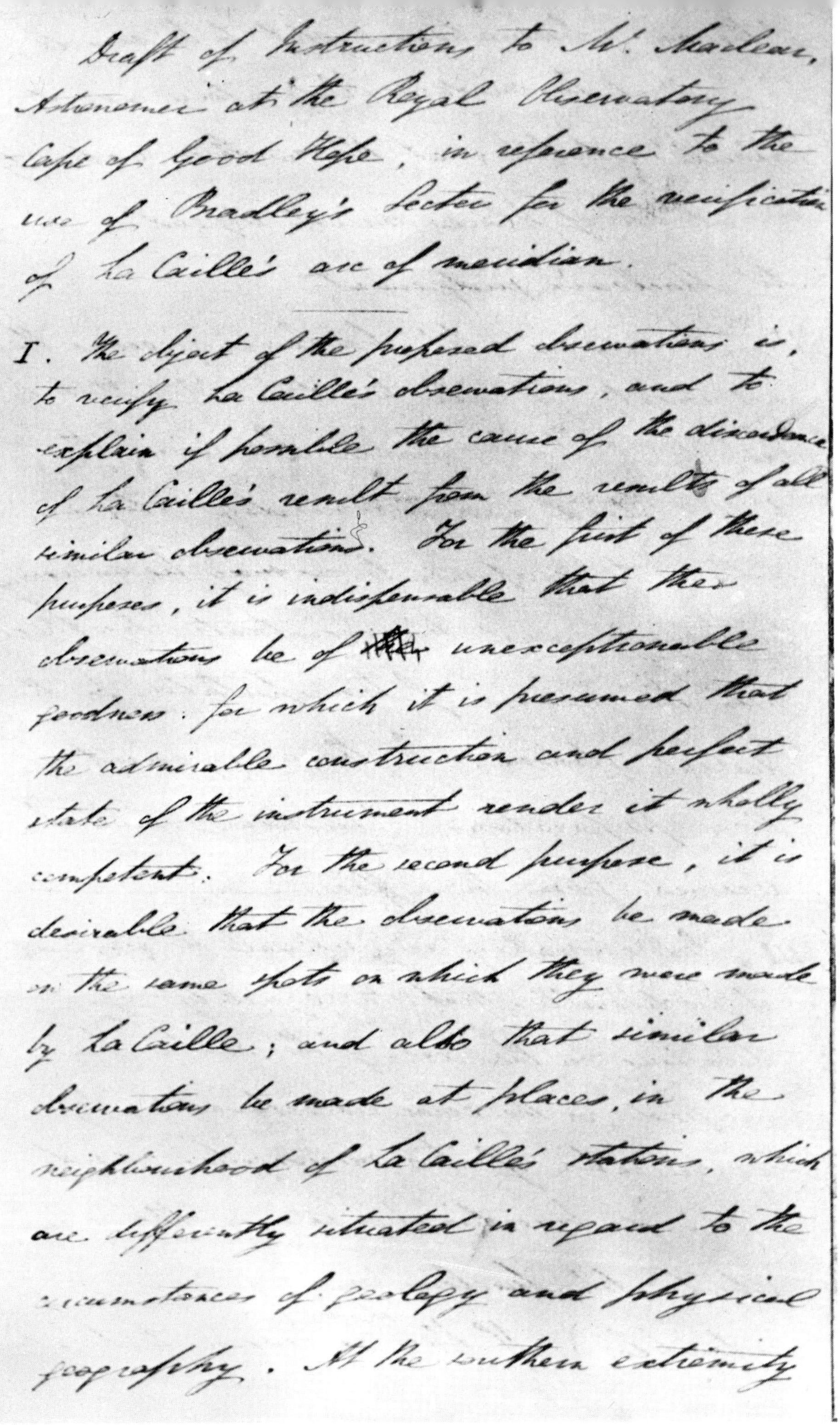

Draft of Instructions to Mr. Maclear, Astronomer at the Royal Observatory Cape of Good Hope, in reference to the use of Bradley's Sector for the verification of La Caille's arc of meridian.

I. The object of the proposed observations is, to verify La Caille's observations, and to explain if possible the cause of the discordance of La Caille's result from the results of all similar observations. For the first of these purposes, it is indispensable that the observations be of ~~[illegible]~~ unexceptionable goodness: for which it is presumed that the admirable construction and perfect state of the instrument render it wholly competent. For the second purpose, it is desirable that the observations be made on the same spots on which they were made by La Caille; and also that similar observations be made at places, in the neighbourhood of La Caille's stations, which are differently situated in regard to the circumstances of geology and physical geography. At the southern extremity

The first page of Sir George Airy's letter of instructions to Maclear.

– Courtesy of Africana Museum, Johannesburg

tion and perfect state of the instrument render it wholly competent. For the second purpose, it is desirable that the observations be made on the same spots on which they were made by La Caille; and also that similar observations be made at places in the neighbourhood of La Caille's stations, which are differently situated in regard to the circumstances of geology and physical geography. At the southern extremity of the arc, observations ought to be made as well as at the modern observatory as at La Caille's station; at the northern extremity the selection of places must depend on Mr Maclear's judgment.

"II. As soon as Mr Maclear receives the instrument, it will be desirable that he provide a tent capable of covering the tripod stand and surrounding stage, with such equipment as may be necessary for himself and his assistants; together with the means of transportation to the distant stations. In the arrangement of these preparations, assistance will be received from the Office . . ."

It is interesting to speculate as to Maclear's feelings when he unpacked the crates containing the instrument. It is possible, of course, that he had seen it before at the Royal Observatory, Greenwich (it is still there, in what is now Flamsteed House). But the "tent capable of covering the tripod stand and surrounding stage" had to be made seventeen feet high, and the colossal instrument, together with its tent and all its trappings, had to lugged around by Maclear on all his subsequent geodetic journeyings. It travelled hundreds of miles by ox wagon over the trackless veld, even as far as the Orange River; and whether it was always possible to observe the injunctions contained in paragraph VII of Airy's letter – "As the instrument is of great historical celebrity, it will be Mr Maclear's duty to preserve it with great care" – is open to the gravest doubt. One pictures the ox wagon containing "the instrument of great historical celebrity" sinking over its disselboom in soft river mud, or slithering to the edge of the precipice on the mountain peak which is now known as the Sector Berg. In any case, Maclear's heart must have been in his mouth so many times that he could well have come to accept this situation as its natural abode!

During his early years at the Cape, Maclear spent a considerable amount of time and effort in trying to locate the exact site of La Caille's observatory, together with the survey points which he had used. That he was not entirely successful was not his fault. La Caille had done his work eighty years before, and since then there had been a great deal of demolition work as well as the erection of new buildings. Under the circumstances, Maclear did well to find some of the points – in one case, the ashes of one of La Caille's signal fires. But on the whole, his only course was to carry out a fresh survey of the same area, fixing new survey points while covering La Caille's original ground as closely as possible.

First, then, Maclear began survey operations on a tract of flat ground near the Observatory. Herschel's rôle in these proceedings seems to have been what Americans would term that of a "kibbitzer" – a person who stands on the sidelines, watching carefully and offering advice. He was certainly helpful. Maclear's measuring-rods were top-heavy, and much of his work was held up when they were blown down or tipped over. Herschel suggested improvements, based on his own experience of such matters, and they were successfully adopted. Unfortunately, the Observatory base-line was washed out by floods. Maclear managed to measure it satisfactorily at a later date.

For the benefit of other surveyors, Maclear then undertook the laying-out and measurement of a base-line on the Grand Parade, Cape Town. In those days of fewer distractions

Measurement of the base line on the Grand Parade in Cape Town, December, 1837.

– Courtesy of South African Archives

his work, performed as it was in full view of the public, seems to have caused a minor riot. The contemporary drawing of the scene of action, given here, will show what is meant; note the policemen with their raised batons! Again the first attempts were unsuccessful, and so the base-line was measured for a second time on 2 December 1837, this time with the painstaking accuracy for which Maclear was already becoming known.

Much to his own frankly expressed satisfaction, Herschel was unable to be present. In a letter to Maclear – it occasionally pleased him to adopt the rôle of effete dilettante, though nobody who knew him was deceived – he wrote: "That *broil* would have done me thoroughly. The next thing I expect to hear is of Messrs. Maclear, Smith, Williams and Co. – ill of brain fevers . . ."

The Smith referred to was that same Charles Piazzi Smyth whom we have already met;

he of the Great Pyramid cult, and future Astronomer Royal for Scotland. At this time he was sixteen years of age, and a tall lad with ability far beyond his years. He was at once accepted, apparently on equal terms, by the Herschel family and the Observatory staff, although the capacity in which he had been sent out was the comparitively humble one of general assistant to Maclear. He quickly proved his worth, and in future undertakings Maclear did not hesitate to entrust him with work of real importance.

After his geodetic preliminaries, Maclear felt ready to embark on the verification and extension of La Caille's arc. The field work, which was exhaustive, started in 1838, and continued until 1847. The results are given in a monumental publication, *Verification and Extension of La Caille's Arc of Meridian at the Cape of Good Hope* (two volumes, London 1866, edited by Sir George Airy, Astronomer Royal). As may be imagined, this is an erudite work, great stretches of which are unintelligible to the layman. Only specialists can view the work properly, and appreciate the effort which went into it; it was of course, an official Cape Observatory publication. Sir David Gill wrote:

"It is impossible to convey within moderate limits an adequate idea of the indomitable energy and perseverance with which this operation was carried out, of the difficulties surmounted, and of the work accomplished within limited means."

To help in this survey, Maclear was loaned the services of a party of sappers from the Royal Engineers at the Castle, under the command of Sergeant John Hemming. Fully conscious of the importance of the undertaking in which he was taking part, Hemming afterwards wrote his own account of the expedition. This throws a revealing light on the social conditions of the time:

"A few casualties occurred to the men on the service. One man was unfortunately drowned while bathing, and another was lost while in a state of intoxication, by a fall into a deep gully. I got a broken head with a bludgeon from a drunken fellow, but providentially soon recovered."

Also in the party were William Mann from the Royal Observatory (about whom more anon), and Charles Piazzi Smyth. Mann also wrote a journal, a typescript of which is now in the South African Library; and Smyth has left a series of exquisitely executed drawings, two of which show a view from La Caille's most northerly station.

Let us pause, then, to picture Maclear and his assistants toiling over terrain of the worst possible kind, the ox wagon containing the monstrous zenith sector an object of constant apprehensive concern; bedevilled by the antics of the sappers; and grimly determined that, come what may, the end result of the expedition would be an accurate re-measurement of La Caille's exasperating arc.

And so it proved. To the undisguised relief of all the astronomers and surveyors of the world, Maclear's results showed that the Earth was indeed the same shape at both ends – oblate, or orange-shaped, instead of being flattened in the northern hemisphere and rounded in the southern, as had been feared. Mountain masses at La Caille's two terminal stations had caused what are termed "discordances", though no fault could be found with the Frenchman's careful observations themselves. Many years later, Maclear's geodetic

View from La Caille's northern point. Sketch by C. Piazzi Smyth, 1842.
– Courtesy of South African Library

work led to the setting-up of the official Government Trigonometrical Survey Office of South Africa.

Meantime, Herschel had been pursuing his programme of cataloging the clusters, nebulæ and double stars of the southern sky. His father, Sir William, had devised a method of observation which he termed "sweeping". This was done by fixing his telescope at a particular declination and recording the objects which were carried through the field of view by virtue of the Earth's rotation. Sir John swept in declination; he was remarkably accurate, and even today some of his observations have not been bettered. True, his catalogue of nebulæ has stood the test of time more successfully than his list of double stars. In many cases he listed optical pairs – that is to say, doubles which are not made up of two stars which are genuinely associated. However, his catalogue of nebulæ was first-class. It even contained several objects which we now know to be external galaxies, many of which are spiral in shape (though the spiral forms were not described until 1845, by the Earl of Rosse – with the aid of his vast 72-inch reflector at Birr Castle in Ireland). Herschel's drawings were, in the main, excellent, but are not consistent; it has often been commented that some easily-observed objects were inaccurately represented, while the much more difficult ones were drawn with surprising correctness.

Another view from La Caille's northern point. Sketch by C. Piazzi Smyth, 1842.
– Courtesy of South African Library

One feature to which he devoted special attention was the nebulæ round the erratic variable star Eta Argûs, now known as Eta Carinæ. (The old constellation of Argo Navis, the Ship, has been split up; Carina is its keel – containing Canopus, the second brightest star in the sky.) Herschel spent many nights in drawing the nebula, and he also made notes about the rise to brilliance of Eta itself. For instance:

"December 19 1837. Eta Argus is now *hardly inferior* to Alpha Centauri, but it *is* inferior. Very like in colour.

"December 20 1837. 1½ a.m. Eta Argus is not so bright as Alpha Centauri but it far exceeds Beta Orionis now at nearly same Alt. It is much nearer to Alpha Centauri than to Beta Orionis (Rigel), Alpha Orionis is small and Procyon trifling compared to it.

"December 22 1837. Eta Argus is far Superior to Rigel (as it was also when all clear & of = Altitudes at an earlier hour) and decidedly exceeds Alpha Centauri – but Alpha is lower by far and there is also haze in sky so comparison not fair – yet I *perceived* no haze in or near Alpha – Eta is now evidently going on to rival Canopus.

"December 23 1837. Eta Argus is larger than Alpha Centauri, and begins now to approach Canopus – There is not so much difference between Canopus & Eta as between Eta and Alpha Centauri."

The station on Sneeukop, Hottentots-Holland. Sketch by C. Piazzi Smyth, 1842. Note tent protecting transit instrument.

– Courtesy of Africana Museum, Johannesburg

At this time Eta Carinæ was probably the most luminous star in the entire Galaxy, but since Herschel's time it has fallen from grace, and for almost a hundred years now it has been invisible to the naked eye. Today it looks very red, and is obviously not a normal kind of star; it is in a class of its own. Astronomers would dearly like to see another outburst. Few accurate estimates of it were made when it was near its peak, so that Herschel's observations were timely.

As a further aside, Herschel was also the last to record Halley's Comet as it receded from the Sun after passing perihelion in 1835.

To carry out his work of cataloging, Herschel had to know the right ascensions and declinations of marker stars which acted as reference points when he wanted to measure the positions of the new objects which he was discovering night after night. In this he was greatly assisted by Maclear, who obtained the required information by using the transit circle and mural circle which had been sent out and set up at the Royal Observatory. Much correspondence between the two men has been preserved – mainly, it is true, on Herschel's side; personal as well as technical matters are included.

Moreover, Sir John kept a diary during his Feldhausen days, and this is certainly one of the most fascinating documents in the whole history of science. Lovers of Africana, as well as those interested in astronomy, will delight in its pages. It is full of self-revelatory passages, and is written in the uninhibited manner of one who has no cause to be otherwise. It must also be said that Herschel's handwriting was absolutely appalling!

That Herschel enjoyed his stay at the Cape is unquestioned. His diary abounds in details of what were, to him, unfamiliar animals, reptiles, insects, trees and flowers which he found on his estate. His wife, who frankly adored him, shared his interests and his enthusiasms. Her almost continual pregnancies must have been a great trial to her, though there is no mention of this aspect of their lives in any of her own or her husband's writings.

Most of the famous men and women who called at the Cape found their way to Feldhausen, their visits being recorded – or not – according to whether the Herschels had anything special to say about them. On Wednesday, 15 June 1836, the young Charles Darwin, then naturalist aboard the *Beagle*, dined with them; the ship's master, Captain Fitzroy, was also among the guests. Herschel recorded the visit briefly in his diary, but Darwin made it the subject of a revealing paragraph in a letter to J. S. Henslow dated 9 July 1836:

"At the Cape Captain Fitzroy and myself enjoyed a memorable piece of good fortune in meeting Sir J. Herschel. We dined at his house and saw him a good few times besides. He was exceedingly good-natured, but his manners at first appeared to me rather awful. He is living in a very comfortable country house surrounded by fir and oak trees, which alone in so open a country, give a most charming air of seclusion and comfort. He appears to find time for everything; he showed us a pretty garden full of Cape bulbs of his own collecting, and I afterwards understood that everything was the work of his own hands . . ."

(One wanders what particular aspect of Sir John's social behaviour struck Darwin as being "rather awful". As far as is known this is the only mention of alleged uncouthness to be found in contemporary records, though one must add that Darwin's statement is softened by the qualifying words "at first".)

On Sunday, 27 November 1836, Sir John Franklin – discoverer of the North-West Passage – called in the evening, together with Lady Franklin, and Herschel showed them the great globular cluster 47 Tucanæ and the Magellanic Clouds. Herschel's diary often refers to social occasions which were linked with "peeps" at the glories of the southern skies: for instance – "Lieut. Williams came to *tea and stars.*"

D'Urban, Napier, Kok and Waterboer – all names in South African history books; they dined, took tea, star-gazed, or simply chatted with the hospitable family. And they all paid return visits.

Herschel's influence on cultural matters while in residence at the Cape was very great. With a small, carefully-chosen committee he laid the foundations of education in South Africa, and was assiduous in ensuring that it was put into effect after he returned to England. When he announced that his astronomical work had been completed, and that he was about to leave the Colony, the South African Literary and Scientific Institution, of which he was President, gave him a specially-designed gold medal which he cherished above many of the other honours which were bestowed on him. A fund was raised to erect

The Herschel Monument as it originally appeared in the grounds of Feldhausen.
– *Courtesy of Africana Museum, Johannesburg*

a monument on the spot where his 20-ft. reflector had stood; and after his departure, this was done. As we have noted, it took the form of an obelisk. It was carved from giant blocks of Craigleath stone from a quarry outside Edinburgh, and it arrived in two packing-cases in August 1841.

Work on the monument took six months to complete. In its foundations were buried glass bottles filled with silver and copper coins; there were also inscription medals and medals of the South African Institution bearing statistical and geographical notices relating to the Colony. There were engravings of nebulæ observed by Herschel at Slough, together with a plan of Maclear's triangulation, connecting Feldhausen with the site of La Caille's observatory and the Royal Observatory, four miles away. Other items, such as details of the discoveries made by Captain Ross in the south polar regions in 1841, and the measurement of the arc of the meridian in 1842, were added before the base of the structure was finally sealed.

The Herschel Obelisk has been criticized as being an ugly memorial to the work of such an æsthetic man. Perhaps it is; but in its setting – especially when its history in known – there is a rightness about it which is strangely moving. It is still to be seen today, in the grounds of the Grove Primary School, which was founded in 1885.

The inscription on the Herschel Monument.

– Courtesy of Africana Museum, Johannesburg

The Herschel Monument during building operations at the Grove Primary School, Claremont.
– *Dr Patrick Moore*

Herschel returned home to a hero's welcome, which seems to have dismayed him. The young Queen Victoria conferred a baronetcy upon him, and his subsequent career was one long succession of triumphs. A contemporary said of him: "Love of truth was in him absolutely untainted by the egotism of the discoverer." Sir John may not have possessed the genius of his father, but his contributions to science covered a far wider range. He lived on until 1871.

Maclear must have missed his friend sorely. Although he had seldom sent written replies to Herschel's copious notes and letters of advice, he had valued them tremendously.

Maclear was indefatigable in the duties of his office. For his work on the remeasurement and extension of La Caille's arc he received the Lalande Prize in 1867 and a royal medal in 1869 (Bradley's zenith sector was returned unscathed to Greenwich in 1850). He originated the Meteorological Commission and also the Commission of Standards for Weights and Measures. He gave valuable help in the establishment of lighthouses in South Africa's coastal waters; he served on the committees of many societies; was a Trustee of the South African Library and Museum – few men have entered so wholeheartedly into the life of their community as did Thomas Maclear. Incidentally, it was he who instructed David Livingstone in the use of the sextant.

Maclear was knighted in 1860, but, sadly, his wife died a year later – a tragedy which, for him, robbed the honour of much of its savour. He retired from the Observatory in 1870, to nearby Mowbray. In 1876 he became totally blind, the worst possible affliction for one who so loved the glory of the stars. He was attended by a devoted family, and retained his interest in public affairs to the last; indeed, his final appearance in public was to welcome H. M. Stanley at a meeting in Cape Town. He died on 14 July 1879. With his wife, he lies buried in the lawn of the Royal Observatory; his grave is a mere fifty yards from that of Fearon Fallows.

For almost the whole of Maclear's régime at the Observatory his principal assistant was William Mann, who proved to be invaluable. Normally, Chief Assistants receive less than their due recognition, and this has certainly been true in the case of Mann, so it seems only right to say something about him here.

On board the Windsor on the Herschels' homeward voyage. A kilted Highlander entertains the passengers by dancing a fling.

– Courtesy of South African Library

The Royal Observatory, Cape Town, in 1865.

– Courtesy of Africana Museum, Johannesburg

He was born in Kent in 1817, and came out to South Africa to join the Cape Staff in 1839, bringing with him various pieces of equipment which had been prepared for re-measuring La Caille's meridian arc. For six years this work occupied him almost exclusively, and he carried out his task with the greatest possible skill. In the process he had to go on long journeys into the wild country of the Clanwilliam district, and sometimes he had to sleep in the open for many weeks at a time. Once, for instance, he went personally to a peak which had been seen over the Karoo country from the Sneeuwkop, and which seemed likely to answer for a connected station; he had only one Hottentot attendant, and before long he ran short of water. In his own words: "Next morning I called a council of war, consisting of myself, to debate on what was to be done. I found there was bamboo enough to last the horses two days, and we two mortals might possibly exist that period on the quantity of bread that remained . . ." He risked an ascent of the mountain, taking with him two rusks to provide lunch, dinner and breakfast; when he reached the summit he had to fill his mouth with pebbles to keep it moist. It is hardly surprising that after experiences of this sort his health was affected, and in 1846 he took a holiday in England.

Transit instrument at the Cape Observatory, 1852.

– Courtesy of Africana Museum, Johannesburg

Maclear wrote: "I feel the loss of Mr Mann's services, especially at the present juncture. His powerful intellect, his unflinching integrity and industry enable me to trust him with confidence on all occasions . . . What he does will be sure to be well done."

Mann returned in 1847, and resumed his routine work. In 1852 he again went to England, this time to bring back the large transit circle which had been made for the Cape, and which was more or less a replica of the famous Airy Transit Circle at Greenwich. On his return, he personally supervised its installation, and indeed carried out much of the work himself.

In 1853 Mann married Caroline, Maclear's second daughter. When, six years later, it was Maclear's turn to spend a period in England, Mann acted as director of the Observatory. By now he had become interested in comets; and it was during a series of long observing sessions studing Encke's periodical comet that he began to suffer from the severe bronchial condition from which he never really recovered. A further trip to England in 1867 brought only temporary relief, and at last, after 1870, he was compelled to resign his official appointment. His health slowly declined, and he died on 30 April 1873.

It would be hard to over-estimate Mann's contribution to the Observatory throughout more than thirty years of service. It may have been unspectacular, but it was none the less valuable for that – and nobody was more conscious of his worth than Maclear himself.

4. THE CAPE: FROM STONE TO SUTHERLAND

MACLEAR's successor at the Cape, Edward James Stone, was a practical astronomer of long experience. For ten years he had been Chief Assistant to Sir George Airy at Greenwich, and had been reducing a mass of observational data with the aim of improving our knowledge of the fundamental constants. His work at the Cape was to be of a similar nature. Many of Maclear's observations still awaited reduction, and Stone applied himself to this task with dedication. He also engaged in systematic observation of the positions of all the southern stars brighter than the seventh magnitude – over 12 000 in all.

It is an interesting fact that almost all H.M. Astronomers at the Cape have carried out extra research pursuits outside the formal duties of their office. Stone was no exception. He gave much thought to the ways in which the speed of sound could be measured, and he found the answer in a manner which was, frankly, lucky. In Cape Town, as in other cities, a noonday gun is fired. The interval between the flash and the report is easily found if the gun is within visual range – as was the case at the Cape Observatory; and Stone made a series of very accurate measurements.

Stone had graduated from King's College, Cambridge, and at an early stage in his career became known and respected for the thoroughness of his investigations. While at the Cape, where he spent nine fruitful years of his working life, he saw the memorable aurora of 4 February 1872. From Klipfontein in Namaqualand he studied the total solar eclipse of 16 April 1874, taking the opportunity to make observations which confirmed Young's spectroscopic discovery of the "reversing layer" above the bright surface of the Sun. On 8 December 1874 he observed the transit of Venus. He afterwards commented that his Cape years were among the most rewarding of his whole life. He returned to England in 1879 upon being nominated Radcliffe Observer at Oxford, and in 1890 produced the Radcliffe Catalogue – which gave the places of 6 424 stars between the celestial equator and declination –25°.

Stone's successor was a Scot, David Gill, who was born in Aberdeen in 1843. He proved to be one of the most outstanding of all the astronomers of South Africa, and it was indeed due to him that in one vital way astronomy as a science entered a new phase.

At the age of fourteen Gill was sent to Dollar Academy. There he boarded with the headmaster, Dr Lindsay, whose interests in mathematics, natural philosophy and chemistry were passed on to his young charge. Gill was a bright, clever, cheery boy with a quality which was to stand him in good stead in later life – a total lack of self-consciousness. His voice was loudly-pitched, with a very pronounced Scottish accent which he never lost. During his days at the Cape a phonograph of the early Edison type was displayed in a shop window in Adderley Street. Intrigued, Gill – who was by this time well-known – entered the shop to examine the instrument, in the company of his wife and several friends. The shopkeeper suggested that he should try it out, and Gill recorded a few sentences. When the recording was played back, Gill's face was a study. He turned to his wife, and said: "Do I r-really r-roll my R's like that?" The burst of laughter which greeted this query not only assured him that the reproduction was faithful, but halted passers-by in their tracks!

Gill's father was a watchmaker with premises in Union Street, Aberdeen. The firm was

Sir David Gill.
– *Courtesy of Africana Museum, Johannesburg*

a distinguished one, holding a Royal Warrant as Watchmakers to Queen Victoria. Gill used to relate how, as a small boy, dressed in a black jacket and waistcoat with silver buttons, a brown winsey kilt and tartan plaid, he and a companion had waited on Banchory station to attend the Queen as she passed through on her way to Balmoral.

After Dollar he spent two years at Marischal College, Aberdeen, as a private student – the term then used to denote those not intending to qualify for a degree. He was a hard worker, but he found time to join in student activities; he became a member of the Volunteer Force, in which he quickly showed himself to be an outstanding marksman.

He became interested in astronomy at a very early age. Indeed, it is difficult to trace the exact time when the "bug bit him", as he himself put it. Certainly it was before his tenth birthday. His younger brother James told the story of how, when the two of them were coming home from a late dance, David pointed upward at a brilliant star and said: "Jem, look at that. Isn't it magnificent?" – to which James replied "Aye . . . aye . . . Come on, Davie – it's three o'clock!"

His father wanted him to join the watchmaking business, and David agreed, though his heart was never in it. For a year he was sent to the manufacturing centre at Besancon, in Switzerland, and it was here that he first showed his exceptional aptitude for making and using precision instruments. He then spent periods in Coventry and in London. In 1863, on his return to Aberdeen, his 74-year-old father made him a junior partner in the firm.

Though he worked conscientiously in his business, at no time did his devotion to astronomy lessen. Indeed, it seemed to grow with the years. In his spare time he helped to rig up an old transit instrument at King's College, Aberdeen, where, with Professor Thomson, he established an accurate time service.

In 1866, in a garden opposite his father's house, he erected a small observatory. In it he installed a telescope with a silvered speculum of 12 inches aperture and 10 ft. focal length. The mounting was of his own making. This telescope, now an historic instrument, found an honoured home in the Calton Hill Observatory many years later.

Using this telescope, with its perfect mounting and excellent definition, Gill was able for the first time to judge his own powers as an observer of close double stars. He had excellent eyesight, and in micrometrical measurement he had few equals. Always a modest man, but having the insight to make an accurate assessment of his own capabilities, he became convinced that his natural gifts should be devoted to astronomy.

On 18 May 1869, at a time when photography was still in its experimental stage, Gill took a picture of the Moon. It can safely be said that it was this achievement – the photograph was an excellent one, even by later standards – which drew attention to the young amateur and started him off on the career which was to bring him international fame, and provide astronomers all over the world with a priceless new research tool.

Thirteen miles from Aberdeen lies Dun Echt, at that time the seat of Lord Lindsay, who was only twenty-four years old but who was already an astronomy enthusiast with a wide knowledge of the subject. Strangely, it was not in Dun Echt or Aberdeen that Lindsay saw Gill's Moon photograph, but in London, at a meeting of the Royal Astronomical Society. He sought out Gill to compliment him, and a friendship began which was to prove life-long.

In those days, before the introduction of inheritance taxes, aristocrats could still afford to use their wealth, if they so wished, to further the arts and sciences. It had long been Lord Lindsay's dream to found an observatory at Dun Echt and equip it with instruments as fine as any in the world. But where to find a Director good enough to satisfy such high standards? It says much for Lindsay's perspicacity that as soon as he met Gill, he realized that his search had come to an end, thirteen miles from his own doorstep. Yet he saw, too, that it might be difficult to persuade the successful young businessman to abandon his clockmaking and accept a salaried post.

He need not have worried. To Gill, Lindsay's tactfully-put suggestion was like an invitation to enter the gates of Paradise. After an excited discussion with his wife – he had married Isabel Black, the highly intellectual daughter of a local farmer, in 1870 – he made arrangements to hand over the conduct of his business to a competent successor, and duly presented himself at Dun Echt.

One of the instruments provided for his use was a 4-inch heliometer. As Gill's future observational measurements were to be made largely with this type of instrument, it may be as well to say something about its structure and use. In a heliometer, the object-glass is cut in half, so that one half may be made to slide past the other, giving a double image of the object under study. It is used to measure very small angular distances as seen through the telescope. Gill's great skill in the use of precision instruments, which was innate and which had been further developed during his watchmaking days at Besancon, helped him to make this device peculiarly his own, and indeed his results remained among the best ever obtained right up to the era of space-probes. After a night spent in using the heliometer, Isabel used to say that he would come into the house shouting and singing "as if he had gone daft"!

In 1874 Gill went to Mauritius, with Lord Lindsay, to observe the transit of Venus. The results were disappointing, although they were made the subject of a Dun Echt publication. Gill next used the 4-inch heliometer to measure the parallax of the minor planet Juno, and he obtained a figure for the Sun's distance which is now known to be very near the true one. A visit to Ascension Island, in which Gill was accompanied by his wife, followed in 1877. There they joined an expedition sponsored by the Royal Astronomical Society for the purpose of observing the close opposition of Mars, measuring the parallax of the planet and thereby deriving a new value for the length of the astronomical unit or Earth-Sun distance. Lord Lindsay loaned Gill the heliometer, and the results obtained were excellent. The estimate for the astronomical unit was much the same as that which Gill had already found from his parallax observations of Juno.

It was while he was living in London, busy with the reduction of his Ascension observations, that news came of his appointment as Her Majesty's Astronomer at the Cape of Good Hope. Appropriately, the first to give him the joyful tidings was Lord Lindsay, who, behind the scenes, had been urging his candidacy. An official letter from the Admiralty followed. But before he left for the southern hemisphere, Gill embarked on an extended tour of foreign observatories, a journey which was to reap a rich harvest for international astronomy in the years to come. He visited – and made friends with – astronomers in Paris, Leiden, Gröningen, Hamburg, Copenhagen, Helsinki, Pulkova and Strasbourg. At last all was ready, and in June 1879 the Gills arrived in Cape Town. Their first view of it as the mists cleared away – the Lion's Head and Rump, and the scattering of white buildings at the blue sea's edge – stayed with them until the end of their lives.

In those days the approach to the Observatory consisted of a rough, muddy cart-track. The grounds were practically a wilderness, and the trees so lovingly planted by Lady Maclear had grown unattended. But beautiful lilies and wild flowers grew everywhere, and the Gills took the place to their hearts.

How they felt when they saw the rooms making up their new home is not on record. Stone, who was already on his way to England, can scarcely be blamed for the fact that all the furniture and fittings had been prepared for auction, and were lying around higgledy-piggledy in the most inconvenient positions. The general impression was far from cosy, especially to two people who had come an immense distance and were looking forward to living there. They stayed at a Cape Town hotel until their rooms at the Observatory were ready, and there – much to their delight – the first person to visit them was Miss Maclear, daughter of the old astronomer.

For a man with Gill's background and ability, the poor instrumentation at the Cape Observatory at that time was frustrating in the extreme. The equipment was outdated, and in need of reconditioning. Luckily, Gill had privately purchased the Dun Echt heliometer from Lord Lindsay, and it arrived in December 1880. In January 1881 a

The Comet of 1882 as seen in South Africa. From the Illustrated London News, 2 December, 1882.

– *Courtesy of Africana Museum, Johannesburg*

young American student named Elkins, who had been studying at Strasbourg, paid Gill what was originally meant to be a "short" visit. In the event he stayed for two years and four months. Using the heliometer, the two astronomers embarked upon the first systematic attack on stellar parallaxes, the results of which proved to be outstandingly successful.

Then came the event which was to lead on to the greatest research programme in Gill's whole career. In 1882 a bright comet appeared in the southern sky. It was brilliant enough to be seen in broad daylight, and in the hours before dawn it added a new glory to the heavens. Gill, of course, was enchanted. He spent many hours in observing it, and only regretted that his friends in England were unable to share his delight. Remembering his Moon photograph of 1869, he fastened a portrait camera to the clock-driven equatorial telescope, and, keeping the cross-wires fixed on the comet's head, exposed the plate for several hours. As he freely admitted, the resulting photograph astounded him. True, he

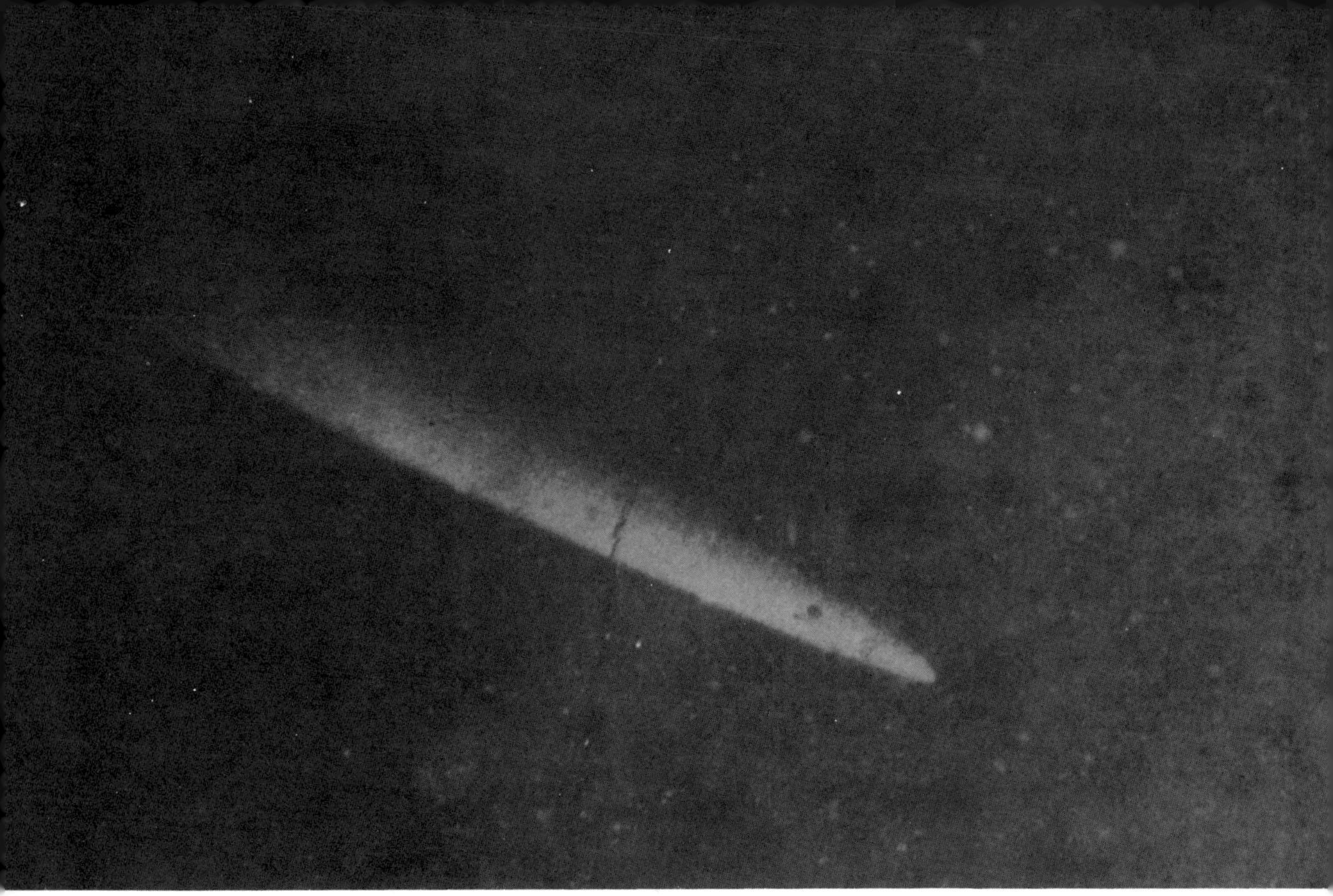

Gill's historic photograph of the 1882 comet.

– Dr Patrick Moore from original source

had expected a good image of the comet, but he had not realized that the background stars would also be shown with absolute clarity and sharpness.

This historic photograph is reproduced here. As Gill studied it, he realized the possibility of photography for making star-maps down to very faint magnitudes. Never a man to sleep on an idea, he wrote a letter to Dallmeyer, the famous optical worker, ordering a larger lens. That letter sailed north on the next mail ship.

The outcome was the famous CPD or *Cape Photographic Dürchmusterung*, which extended the *Bonn Dürchmusterung* down to the south pole of the sky. Funds for the work were sought from the Royal Greenwich Observatory, but were refused, mainly because of a misunderstanding. Fortunately, the Royal Society gave him a grant which covered the expenses of bringing a specialist photographer out from England, and a generous friend donated a further sum. There was still a shortfall, however, so Gill made it up out of his own pocket – the not-inconsiderable amount of £350 per year (worth many times as much today), during all the years that work on the *Dürchmusterung* was in progress. The finished catalogue gives the brightnesses and approximate positions of nearly half a million southern stars.

1. The Devil's Peak from the Observatory
2. The Heliometer Observatory
3. Front of the Royal Observatory
4. The Repsold Heliometer
5. The Old Wind Tower, and Photographer's Temporary Dark-room
6. The Zenith Telescope, or Bent Transit Instrument
7. The Great Indian Theodolite
8. The Astronomer at his Desk
9. The Transit Instrument
10. Theodolite and Zenith Telescope Huts
11. Stellar Photographic Instruments

SKETCHES AT THE ROYAL OBSERVATORY, CAPE TOWN

Sketches from *The Graphic*, August 27th., 1892.

– *Courtesy of Africana Museum, Johannesburg*

The Victoria Telescope, Royal Observatory, Cape of Good Hope.

– *Courtesy of Africana Museum, Johannesburg*

Even while work on the CPD was going on, a larger and even more ambitious project was put in hand by the Paris Observatory. This was nothing less than a photographic chart of the whole sky, showing stars down to the fourteenth magnitude. In addition there was to be a catalogue giving the precise positions of all stars down to the eleventh magnitude – more than a million stars altogether.

A scheme of this kind could be carried through only with the co-operation of many astronomers and many observatories. As the acknowledged "father of astrography", Gill was prominent both in the organizing and the carrying-out of this great work. His many friendships among astronomers, made or strengthened during his "grand tour" before he had set out for the Cape, now stood him – and the *Carte du Ciel*, as the project was called – in good stead. A major portion of the work was carried out at the Cape Observatory, and a fine new telescope, the astrographic refractor, was acquired. Yet in spite of the devotion of Gill and his colleagues, work on the *Carte du Ciel* was not completed until a few years ago. The whole concept was simply too big for late-19th century facilities.

The astrographic refractor proved to be a lucky instrument, for it was the direct cause of the Cape Observatory's acquiring a first-rate telescope of twice that size. Frank McClean, a Fellow of the Royal Society, a British engineer and enthusiastic astronomical amateur, used the astrographic on many occasions. Some years later he offered the Observatory, as a gift, what was to become known as the Victoria Telescope, which was completed in 1901. Yet another fine instrument added not long before Gill made his final bow as Director was a new reversible transit circle. It was designed by Gill himself, and has become the pattern for most of the transit circles which have been made since.

Gill was knighted on 24 May 1900. In 1906, because their health was failing, Sir David and Lady Gill retired to 34 De Vere Gardens, Kensington in London. There they lived happily – and actively – during their remaining years. Gill died on 24 January 1914, and was buried in a grave he had chosen for himself in Aberdeen.

He left his successor what had come to be one of the finest and best-equipped observatories in the world. Moreover, his position in the Imperial (as distinguished from the Colonial) Civil Service had enabled him to exert unusual influence upon affairs in the South Africa of his day – a position which he invariably used to the Colony's advantage.

Unlike his predecessors, Sydney Samuel Hough, who followed Gill as H.M. Astronomer at the Cape, already knew the Observatory well by the time he was appointed. Indeed, he had worked there as Gill's assistant for the previous nine years. Born in Stoke Newington, in London, in 1870, he was educated at Christ's Hospital and at St John's College, Cambridge, of which he became a Fellow. Although he was an authority on celestial dynamics and was a brilliant mathematician, Hough's special interest was in the dynamical theory of tides. When he arrived at the Cape he had had little or no experience in practical astronomy, but he learned quickly, and Gill came to place complete confidence in him.

Hough's principal task was to ensure that Gill's programmes were completed, and he accomplished it brilliantly. Five of the twelve volumes giving the results of the measures for the Cape Zone of the Astrographic Catalogue were published during his term of office. With his Chief Assistant, J. K. E. Halm, he originated much of the southern-hemisphere work on the proper motions of stars. It has been said of Hough that his period as Director was not one of change, but of solid achievement.

Staff of Royal Observatory, Cape of Good Hope, 1906. Sir David Gill (centre) flanked by his successor S. S. Hough (dark moustache), and J. Lunt (beard).
– *Courtesy of Africana Museum, Johannesburg*

Unfortunately, he had the bad luck to be H.M. Astronomer at a time when His Majesty's Forces were engaged in the bitterest struggle which had ever been waged up to that time. He used to complain that instead of watching stars, most of his staff were watching starshells in Flanders. And when the 1914–1918 war came to an end it was replaced as a world-scourge by the great influenza epidemic. Sadly, Hough's wife was one of the first victims. Hough himself contracted his final illness while still in charge at the Cape. He died in England in 1923.

Hough is remembered personally today by a dwindling few. Harold Spencer Jones, who followed him, is well and happily remembered by a great many people, including one of the authors of the present book.

Spencer Jones came to the Cape from the Royal Greenwich Observatory, where he had been senior Chief Assistant. Born in London, in 1890, he went to school in Hammersmith and then on to Jesus College, Cambridge, of which he became a Fellow. His terms of office

Halley's Comet as seen from Port Elizabeth.

– Courtesy of Africana Museum, Johannesburg

at the Cape – 1923–1933 – included the years of depression; and as these coincided with the need to renew and repair equipment which could not be properly maintained during the war years, he had a continuous struggle to obtain the essential funds. Fortunately, he was a convincing and experienced advocate who enjoyed the respect of the authorities in all departments, so that his appeals did not go unheard. The Observatory accommodation was modernized, and a new office block was built during his time at the Cape, while in the field of research the backlog of old programmes was cleared away and new projects started.

Most of the members of the Observatory staff had been engaged by Gill, and were nearing retirement age. It was Spencer Jones who initiated the policy of replacing retired staff members with fully-qualified South Africans whenever possible, a policy which has been continued with marked success. Indeed, a later Director remarked, mournfully, that the young South African astronomers were so good that many of them were tempted overseas when they had gained experience.

The Cape Astrographic Zone, photographed by Gill and his staff for the *Carte du Ciel*, was re-photographed in its entirety, so that the new plates could be compared with the old ones. From these comparisons, the proper motions of more than 40,000 stars were measured. The Victoria Telescope, which had been used in a radial velocity programme, was devoted to a photographic parallax programme for the purpose of measuring the distance of several hundred of the closer southern stars.

At the end of 1930 the minor planet Eros made an exceptionally close approach. Until well into 1931, sixteen northern and six southern observatories combined in a project, largely organized by Spencer Jones, to find the parallax of Eros and thence to make a new measurement of the astronomical unit. Much of the work was undertaken at the Cape, and it was Spencer Jones who made the resulting calculations. His derived value for the unit was rather over 93 000 000 miles. This is now known to be rather too large – the latest value, from radar measurements, is 92 957 209 miles – but it nevertheless represented a notable piece of research. However, the greatest programme which was set in motion during Spencer Jones' régime was the Cape Photographic Catalogue for 1950, giving the precise positions, colours, magnitudes, and – with a few exceptions – proper motion for 70 000 southern stars.

Spencer Jones returned to England in 1933, to a knighthood and to the post of Astronomer Royal. He paid one return visit to the Cape, in 1948, to arrange a co-operative research scheme with the Radcliffe Observatory. His period as Astronomer Royal, which lasted until 1954, was memorable in every respect – and saw the virtual completion of the move of the main instruments from Greenwich Park to Herstmonceux. Spencer Jones died in 1960, working actively on his research programme until a day or so before his death.

John Jackson, a Scot from Paisley who had been educated at Glasgow University and at Trinity College, Cambridge, took over the reins from Spencer Jones. He, too, had been senior Chief Assistant at Greenwich; and like Hough, he was to suffer the staff disruption caused by years of war. In spite of this, his term of office was noted for its production of a mass of valuable work. Jackson's special concern was the stellar parallax programme; he undertook the necessary reductions and observations personally, at a time when two-thirds of his staff were "up north". Jackson retired in 1950, having measured the distances of over 1 600 southern stars with greatly improved accuracy and having advanced or completed all the major projects which had been going on at the time of his appointment. He was succeeded by Richard Hugh Stoy, who proved to be the last holder of the title of "Her Majesty's Astronomer at the Cape of Good Hope".

Stoy was born at Wolverhampton in 1910, and educated at Wolverhampton Grammar School and at Gonville, after which he went to Caius College, Cambridge. (Six out of the nine Cape Astronomers have been Cambridge men.) A Commonwealth Fellowship at the Lick Observatory, California, followed. There he stayed for two years, making a special study of planetary nebulæ. At the age of twenty-five he became Jackson's Chief Assistant, thus beginning an association with the Cape Observatory which was to last for thirty-three years.

Stoy's personal contributions to astronomy have been outstanding. He it was who initiated the photographic photometry programme which has led to the determination of the colours and magnitudes of the 70 000 stars in the CPC. His meticulous work in the establishing of photometric standards grew into the wide-ranging programmes of the 1950s and 1960s devoted to precision photoelectric photometry.

It was also Stoy who, when a Department of Astronomy was created at the University of Cape Town, in 1958, became its Honorary Professor. Today the Department is extremely active, and more will be said about it below. But, of course, Stoy's main work was connected with the Cape Observatory itself, and great changes took place during his régime, which began in 1950 and ended in 1968. Whenever possible, renovations to the equipment were put in hand; the Gill Transit Circle, the Victoria Telescope and the astrographic telescopes were all overhauled. An 18-inch reflector, intended mainly for photoelectric use, was bought and fitted to the mounting of the Gill heliometer. Then there was the 30-inch reflector, presented to the Observatory by Dr W. H. Steavenson, which proved to be a valuable asset, and was used chiefly for infra-red photometry.

Steavenson was a remarkable man. His doctorate was a medical one, and he always ranked himself as an amateur in astronomy; nevertheless, he rose to become President of the Royal Astronomical Society, and it was generally agreed that his knowledge of optics and of observational techniques was second to none. (His unmatched skill as an observer was all the more remarkable because almost all his life he was blind in one eye.) It was he who carried out the surveys which led to the selection of Pretoria as a site for the Radcliffe Observatory, and the decision was perfectly correct at the time; neither Steavenson nor anyone else could forsee how the spread of Pretoria itself would interfere with the seeing conditions in later years. Steavenson was intimately connected with the British Astronomical Association, which was (and is) largely amateur but which has strong professional connections – together with an observational record which is probably unrivalled. Steavenson served a term as President – so, for that matter, did Spencer Jones – and was at various times Director of its Mars Section and its Instruments and Observing Methods Section. He was also closely associated with Cambridge, and it was there that he set up a 30-inch reflector of fine quality. When he grew older, and realized that his career as an active observer was coming to an end, he arranged for the transfer of the telescope to Cape Town so that it could continue to be used to its maximum capacity. This was typical of Steavenson, whose generosity and kindliness were without limit.*

However, the most important addition to the equipment during Stoy's Directorship was a 40-inch reflector, which was completed and installed in 1963. It was named the Elizabeth Telescope because it first received approval "in principle" in 1953, the year in which Queen Elizabeth II was crowned. There were also many other instruments, and the whole equipment at the Cape was thoroughly modernized. Stoy himself was the driving force; the Observatory can never have had a more effective Director, or one who was more universally popular.

By 1968 some of the major programmes were almost complete, and there were no

*I have good reason to know this. When I was a boy of eleven, I joined the British Astronomical Association, and there came the all-important question of choosing a telescope. Steavenson, then one of Britain's most famous astronomers, took the trouble to come all the way to my home in East Grinstead and select for me a 3-inch refractor, which I still have and with which I published my first lunar papers. Steavenson remained on hand to help and advise, right up to the time of his death in 1975. I was only one of many to receive help from him. – P.M.

The Main Building at the Cape Observatory. Dr David S. Evans and Dr Patrick Moore walking towards camera.

– *Pete Collins*

arrears of publication, a state of affairs which had been almost unknown since the start of serious observing in 1829; in November 1968 Stoy retired, to return to Scotland as Deputy Director of the Royal Observatory at Edinburgh and Honorary Professor of Astronomy at Edinburgh University. His Chief Assistant, Dr D. S. Evans, had already left to become a Professor of Astronomy at the University of Texas. Meanwhile, events had taken a new turn.

Like all cities, Cape Town was expanding. Growth means artificial lighting, and the proximity of some major football stadia did not help. The sky was no longer dark and transparent, and the Observatory was faced with a problem similar to that at Greenwich during Spencer Jones' spell as Astronomer Royal. Moving a city is no easy matter, and in the modern age shifting a football ground is well-nigh impossible. The only alternative was to move the Observatory – or, rather, its practical equipment. As we have noted, the Royal Greenwich Observatory had already been shifted to the relatively unpolluted skies of Herstmonceux Castle in Sussex – a transfer which was started in Spencer Jones' time and was completed during the régime of Sir Richard Woolley, who also comes very much into the South African story. The Cape Observatory authorities came to a similar decision.

Where should the new station be placed? There was no thought of dismantling the

The Old Library at the Cape Observatory.

– Dr Patrick Moore

Old Cape Observatory itself, but it was a clear albeit unpalatable fact that the original site was no longer satisfactory. Tests were made in all sorts of locations, and eventually the choice fell upon Sutherland, on the border of the winter and summer rainfall areas. With its altitude of over 5 000 feet, it is one of the coldest places in the Republic. It also has one of the highest percentages of clear nights with rock-steady seeing conditions.

When the final decision was made, it was courageous in the extreme – and also, it must be said, controversial. Not only would the 40-inch Elizabeth telescope and other Cape instruments be moved to Sutherland, but also some of the main equipment from other South African observatories, notably the 74-inch reflector from Pretoria and the 20-inch reflector from Johannesburg. This would, inevitably, mean the end of some historic institutions. The Radcliffe Observatory at Pretoria was doomed; the Republic Observatory at Johannesburg was to all intents and purposes relegated to a rôle of professional inactivity; the Cape Observatory itself would become essentially an administrative headquarters.

Arguments raged, and some of them became decidedly heated, particularly insofar as Johannesburg was concerned (as we will see later). Boyden Observatory was unaffected, because the conditions at Bloemfontein are strictly comparable with those at Sutherland, but with regard to Pretoria and to the Cape itself the decision was logical by any standards; if South African astronomy were to be organized with the maximum possible efficiency,

The telescope domes at Sutherland Observatory.

– Dr Hamish Campbell

the best telescopes would have to be taken to the best observing site. So Stoy was the last H.M. Astronomer; when he retired, G. A. Harding, of the Royal Greenwich Observatory, came as Officer-in-Charge, and remained until the transfer was complete. Harding was no stranger to South Africa, since he had previously spent some years at the Radcliffe Observatory. Then, at the start of the new era, Sir Richard Woolley arrived to become the first Director of the South African Observatories.

Certainly there was nobody more qualified to take charge than Richard van der Riet Woolley. A Cape graduate, he had been outstanding at Greenwich and then as Director of the Commonwealth Observatory in Australia before returning to England in 1956 as Astronomer Royal – the last man, incidentally, to combine this post with that of the Director of Greenwich Observatory. (On his retirement, at the end of 1971, Dr Margaret Burbidge succeeded him as Director for a brief period, while Sir Martin Ryle, the great Cambridge radio astronomer, became Astronomer Royal.) As well as being a first-class administrator, Woolley also had an unsurpassed reputation as an observational astronomer, and he was ideally suited to the task of setting up the new Sutherland station.*

*As an aside, it may be recorded that Sir Richard Woolley has something in common with Edmund Neison, the only Director of the old Natal observatory – he is a very strong tennis-player; and like Sir Bernard Lovell, he is also an excellent cricketer!

Sir Richard van der Riet Woolley at Sutherland.

– Dr Patrick Moore

Shifting the instruments to Sutherland was, naturally, a major problem, but it was overcome within a few years. The Sutherland station was officially opened on 15 March 1973. The ceremony was performed by the Prime Minister, Mr Vorster; among those present were Dr Brink (President of the CSIR or Council for Scientific and Industrial Research), Mrs Margaret Thatcher (then the British Secretary of State for Education) and, of course, the new Director, Sir Richard Woolley.

The Elizabeth reflector was moved in its entirety, and was given a new dome. It was put into use at once, and was fitted with a spectrograph borrowed from Greenwich; it was also used for direct stellar photography, and soon showed that its performance was much better at Sutherland than it had been at the Cape. Then there was the so-called MRM or Multiple Refractor Mounting, which had been obtained from the English firm Grubb Parsons in 1964, and which also came from the Cape. As the name suggests, several telescopes are carried on the same mount; the main refractor has a very wide field, and Woolley has himself been making extensive observations with it, mainly for variable star research. The MRM, too, has a new dome; so has the third of the original Sutherland telescopes – the 20-inch reflector which had long been installed at the Republic Observatory in Johannesburg.

The Official Opening of the Sutherland Observatory, 15 March, 1973. Mrs Margaret Thatcher, Minister of Education and Science in the British Conservative Government, addresses the congregation.

– *Jack Bennett*

All these were in position, and in full use, by 1974. Next came the 74-inch reflector from the Radcliffe Observatory in Pretoria, which remains, so far, the largest telescope in South Africa. The present writers remember seeing the skeleton tube lying in the open at the beginning of 1975, looking rather forlorn beside its half-completed dome; but it too was installed by 1976, and it has already begun a new lease of life.

Sutherland itself is a remote place; the town is small, so that there are no inconvenient artificial lights (and in any case the Observatory is several miles away). Driving to Sutherland from Cape Town means a journey of several hours across country which is fascinating even if somewhat bleak – the Karoo has a magic all its own. The usual procedure is for an astronomer to come to Sutherland for a week or two, carry out observational work, and then go back to the Cape itself to analyze and evaluate the results. Not that Sutherland itself is uncomfortable; far from it – and it serves its purpose well. Even at this comparatively early stage, it is safe to say that the venture has been a success.

A scene at the Official Opening of the Sutherland Observatory.

– Jack Bennett

The 74-inch Telescope Building at Sutherland seen between the 40-inch and MRM Domes.
– Dr Hamish Campbell

Since the story of the Cape Observatory is so vital in South African astronomy, it may be useful to list the Directors and Chief Assistants:

Directors

1820–1831 Rev. Fearon Fallows
1831–1833 Thomas Henderson
1833–1870 Sir Thomas Maclear
1870–1879 Edward James Stone
1879–1907 Sir David Gill
1907–1923 Sydney Samuel Hough
1923–1933 Sir Harold Spencer Jones
1933–1950 John Jackson
1950–1968 Richard Hugh Stoy
1969–1973 G. A. Harding (Officer in Charge)

Chief Assistants

1835–1845 Charles Piazzi Smyth
1847–1872 William Mann
1873–1897 W. H. Finley
1898–1907 Sydney Samuel Hough
1907–1927 J. K. E. Halm
1927–1935 H. Horrocks
1935–1950 Richard Hugh Stoy
1950–1968 David S. Evans

Sir Richard Woolley is the first and, so far, the only Director of the combined South African Observatories. He assumed office in 1973.

One important aspect of the new arrangement is that it is international, and came about as a result of an agreement between the London Science Research Council and the South African CSIR. Up to now, at least, politics have not interfered; one can only hope that this state of affairs will continue, though, naturally, there is an increasing tendency for South African-born astronomers to take leading rôles.

All in all, it is fair to say that the Cape has been the main centre of South African astronomy from the earliest days right up to the present time. The Cape Observatory itself is steeped in history, and it must surely be the most picturesque observatory in the world, with Table Mountain in the background and the sea close by; the buildings themselves are impressive, and the Library is beautiful. Sutherland, stark and isolated though it may appear at first sight, is established as one of the most important astronomical observatories in the world. No doubt further equipment will be added in the future, and it may be fitting to end this account with a quote from Sir Richard Woolley himself, made to one of the present writers in 1975. "People who run theatrical shows are always looking for what they call angels," he said when we were carrying out a television broadcast. "Well – if anyone now viewing this programme would like to give me a couple of million rands, I could use it!"

5. THE REPUBLIC OBSERVATORY: TRIUMPH AND CONTROVERSY

UP TO THE end of the last century, the astronomy of South Africa was to all intents and purposes the astronomy of the Cape. It was here that Father Tachard had landed; here that La Caille and Herschel had worked, and here that the nation's first major observatory had been set up, with its succession of distinguished Directors beginning with Fearon Fallows. From being a "Dismal Swamp", to use Henderson's unflattering term (though one can well appreciate his dislike of companions as uncongenial as snakes and hippopotami), the Cape Observatory had become world-famous, and as the nineteenth century drew to its close the Director was Sir David Gill, always a champion of the excellence of South African skies. But what about Johannesburg, a city growing daily in importance? There was no observatory in the Transvaal, where the climate is superb; and in 1902 the South African Association for the Advancement of Science decided to take action. They petitioned the Transvaal Government for the establishment of a meteorological and astronomical observatory, and their plea fell on receptive ears – at least to some extent.

Meteorology was the first concern. An observatory was duly set up (a photograph of it is shown here), and came into official existence, as the Government Meteorological Observatory, on 17 January 1905, though in fact it had been operating since 1 April 1903. The opening ceremony was performed by Lord Milner, who had openly expressed the view that Transvaal farmers needed weather forecasts rather than stars. But even at that ceremony, Lord Milner may have guessed that a change would not be long delayed, because of the choice of the first Director of the Observatory: Robert Thorburn Ayton Innes.

Because Innes looms so large in the story of astronomy, not only of the Transvaal but of the world, it seems necessary to say something about his early career, which was extraordinary by any standards. Had Innes been a different kind of man, the Transvaal Observatory would have become a different kind of place.

Innes was born on 10 November 1861 in Edinburgh, so that he was a Scot through and through. He was the eldest of a dozen children, and left school at the age of twelve. Thereafter he was entirely self-taught, which was in some ways a good thing inasmuch as he was able to bring a refreshingly unconventional attitude to all his work in later years. (He was unconventional in other ways also. In middle age, he decided that the Transvaal heat mitigated against wearing a tie, so that he would do so no more – and he never did. Even when presented at the Dutch Court he declined to put one on; he must have looked unusual in official tails but minus any neck garment!) During his teens he showed that he was a mathematician of quite exceptional ability, and he published some papers on celestial mechanics which caused his seniors to raise their eyebrows markedly. At the age of seventeen he became a Fellow of the Royal Astronomical Society, which was a rare honour for one so young.

Yet he had no degree (he did not acquire one until 1923, when the University of Leiden conferred an honorary doctorate on him), and there was the problem of earning a living, particularly as in 1884 he had married; his wife was Anne Elizabeth, née Fennell, by whom he had three sons. What should he do? He had one offer from America, another

The Official Invitation to the Opening of the Government Meteorological Observatory, 17 January, 1905.

– Courtesy of Africana Museum, Johannesburg

from Australia – neither in the field of astronomy. He accepted the Australian invitation, and with his wife he emigrated to Sydney, where he became a wine-merchant. No doubt he was a very good wine-merchant, and the financial problems were pleasingly solved; but his heart was in the skies. He was encouraged by several Australian astronomers, among them W. F. Gale, celebrated as an observer of comets (a periodical comet which he discovered is named after him). Gale loaned him an old $6\frac{3}{4}$-inch Cooke refractor, and Innes began his real work: the study of double stars in the southern hemisphere.

Double stars are of unique interest in astronomy. Rather surprisingly, most of the stellar pairs are physically-associated systems or "binaries"; optical doubles, in which one component is much more remote than the other and merely happens to lie in much the same

Formal Opening of the Government Meteorological Observatory by Viscount Milner, High Commissioner and Governor.

– Courtesy of Africana Museum, Johannesburg

direction, are relatively rare (though, of course, they do exist). Binaries are of various kinds. Some are so widely separated that the components take hundreds or even thousands of years to complete one revolution round their common centre of gravity. Others have shorter periods; for instance Alpha Centauri is made up of two components, one rather superior to the other, with a revolution period of 80 years. (Alpha Centauri is of special importance in South African astronomy. As we have noted, it was the star whose distance was measured by Henderson from the Cape, and later on Innes too was to be concerned with it.) Then there are binaries with really short periods – and some whose components are so close together that no telescope will split them; they have to be detected by less obvious methods.

Binary systems make it possible to measure the masses of the components, a problem

The Government Meteorological Observatory in Johannesburg in 1905.
– Courtesy of Africana Museum, Johannesburg

which is very much more difficult with single stars. Innes, of course, knew this; but before the theorists can come into the picture, the doubles themselves have to be studied – not once only, but over periods of many decades, so that the relative movements can be measured and the orbital characteristics found. First, Innes began to search for new doubles. After a mere thirty hours' search with the $6\frac{3}{4}$-inch refractor he had discovered twenty-six. He published his original list in December 1894; he acquired a larger reflector, and soon issued a second list of new pairs. Clearly, the field of discovery was wide open.

Wine is all very well in its way, but Innes had no wish to make it his permanent career. Now that he had some published papers to back his claims he wrote to Gill at the Cape, offering his services. Gill was impressed, but he was limited in the numbers of staff members he could employ, and all he could suggest was a relatively humble post at a small salary. If Innes cared to throw up his promising Australian connections and come to the Cape as secretary-cum-librarian-cum-accountant, he could do so. Innes, being Innes, accepted, and in 1896 he arrived in South Africa, ready to devote the rest of his life to astronomy.

At first his personal research was confined to what was officially his free time, but even so he managed to accomplish a great deal. By 1898 he had discovered another 280 double stars, and in the following year he published his first *Reference Catalogue of Southern Double Stars*, dealing with 2 140 pairs. He also concentrated upon variable stars, and in addition he measured the proper motions of many stars in the southern sky. He even revised the *Cape Photographic Dürchmusterung*, one of the most important of existing star-catalogues.

Robert Thorburn Ayton Innes. The first Union Astronomer.

– Courtesy of South African Library

Gill was even more impressed. Then, in 1903, came a request for him to recommend a Director for the new Transvaal Observatory. True, it was meteorological; but who better than Innes? There can be no doubt that Gill knew what Innes' plans would be, and he approved of them. So Innes left for Johannesburg, and in the midst of his meteorology he lost no time in laying the foundations of a proper astronomical establishment.

He could not move too quickly, and until 1907 astronomy had to play a minor rôle (though one must mention the visit of a Russian astronomer, Backlund, in 1905, at whose instigation the Transvaal Observatory was authorized to play a part in an international scheme for determining the variation in latitude – using special instruments loaned by the Russian Government). But at last, in 1907, Innes felt able to make a move in what he regarded as the right direction. Aided and abetted by Gill, and with the help also of Dr Theodore Reunart of Johannesburg, he obtained a 9-inch refractor, and the transition was well and truly under way. The first recorded observation with the Reunart telescope was made on 3 October 1907, to note the apparent disappearance of Saturn's rings as the

Installation of the $26\frac{1}{2}$-inch telescope at the Union Observatory, Johannesburg, 1925.
– *Courtesy of Africana Museum, Johannesburg*

Earth passed through the plane of the ring-system. Innes then began his classic series of observations of the phenomena of the four large satellites of Jupiter. But the Reunart telescope was not large enough to satisfy him, and he yearned for something more powerful.

Events moved more rapidly than he had dared to hope. At his instigation the Government was persuaded to place an order for a really large telescope – a refractor with an object-glass $26\frac{1}{2}$ inches in diameter. In 1909 the lens was ordered from the Dublin firm of Sir Howard Grubb. Unfortunately there were seemingly endless delays; casting a lens of this size is no easy matter, and there were other troubles also, so that the refractor was not finally installed until 1925. Innes himself had only two years' use of it, but this was the sole frustration in an otherwise triumphant career.

Also in 1909 the Observatory had a stroke of good fortune, in the shape of the gift of a fine 10-inch photographic refractor. It was presented by a wealthy English amateur named John Franklin-Adams, who had already earned himself a considerable reputation for his personal work. Franklin-Adams was born in London on 5 August 1843, and became a businessman; apparently he took virtually no interest in astronomy until he was 47 years

The Franklin-Adams Telescope at the Union Observatory. R. T. A. Innes (in hat) and H. E. Wood.

– Courtesy of Africana Museum, Johannesburg

old, but he then proceeded to make up for lost time. Even today, the Franklin-Adams photographic maps of the southern sky are regarded as standards.

Just before his death, in 1912, Franklin-Adams gave the Transvaal Observatory another instrument – the so-called twin telescope, consisting of a 6-inch photovisual refractor together with a 7-inch visual refractor. Both it and the 10-inch were ideally suited for mapping projects, as well as for the discovery and observation of those interesting little worlds which we call the minor planets or asteroids.

Next there came a major change in organization. The Transvaal was one of the four states which merged to make up the Union of South Africa; their meteorological services were combined into the Union Meteorological Department, with its headquarters in Pretoria; and the Transvaal Observatory, renamed the Union Observatory, was free to devote its whole time to astronomy. The first of Innes' ambitions had been realized.

The official change dated from 1 April 1912, and Innes became the first Union Astronomer; it is said that he is the only man who has been transformed from a meteorologist

Interior of observing hut of Franklin-Adams Telescope, Union Observatory.
– *Courtesy of Africana Museum, Johannesburg*

into an astronomer by an Act of Parliament! Since there were, in all, only four Union Astronomers, it may be worth while to list them without further ado:

1912–1927, R. T. A. Innes
1927–1941, H. E. Wood
1941-1956, W. H. van den Bos
1957–1965, W. S. Finsen

All were distinguished; all carried out sterling work with the ever-increasing facilities available at the Observatory. Innes himself was indefatigable. Though double stars remained his premier concern, he also made very important observations of minor planets; he studied comets, improved the design of astronomical equipment, and undertook sig-

The four successive Union Astronomers: R. T. A. Innes (left) 1912–1927; H. E. Wood (right) 1927–1941; W. H. van den Bos (centre) 1941–1956; W. S. Finsen (at desk) 1957–1965.

The Old Government Meteorological Observatory, Johannesburg.

– Dr Patrick Moore

nificant mathematical researches. Indeed, he was primarily a mathematician, though he is best remembered today for his outstanding observational results. One of his most famous discoveries was that of Proxima Centauri, the faint companion of Alpha Centauri, which was subsequently found to be the nearest star beyond the Sun; it lies at a distance of a mere 4.2 light-years. Innes did not happen upon it by sheer chance. He thought that there might be a dim third star in the Alpha Centauri system, so he started a systematic search for it; he took numerous photographs – and in 1912 he found it. Around this time, too, he advanced the use of the blink-microscope, which is a device for examining exposed photographic plates. In essence, the two plates to be compared are placed side by side in the instrument, and each is observed in turn; by flicking to and fro in rapid succession, a moving object will be detected with relative ease. Proxima was located in this way, and so, subsequently, were many minor planets.

Innes was extremely far-sighted, and he wanted to make the maximum use of the Observatory equipment. Mainly because of his friendship with Willem de Sitter, one of the great pioneers of relativity theory, he forged a close link with the Leiden Observatory in Holland, and visitors from Leiden became very regular. One of them was Willem van den Bos, who came in 1925 on a temporary basis but who never went back to Holland – and subsequently became Director at Johannesburg. The Leiden connection had very important results in after-years, and led to the setting-up of the outstation at Hartebeespoort, of which more anon.

In 1927, the date of publication of his second and last catalogue of southern double stars and two years after the great 26½-inch refractor was put into commission, Innes finally retired from the Directorship. He left behind an impressive record. He is credited with the discovery of 1628 new double stars, and he made many thousands of measurements; his asteroid work was equally outstanding – for instance, in 1924 he suspected that the shape of the strange minor planet Eros was unusual, and using the great refractor in 1931 his successors, van den Bos and Finsen, confirmed this, managing to measure the rotation period. (Eros was discovered by Witt, at Berlin, in 1898. It has an orbit which swings it inside that of the Earth, and sometimes, as in 1931 and 1975, it can come well within twenty million miles of us. The former approach was used to help in measuring the length of the astronomical unit, as we have already noted; and in this work Union Observatory astronomers played an important part. We now know that Eros is shaped rather like a cosmical sausage, 18 miles long but only 9 wide at its thickest part.)

Innes was a man of many interests. He was, for example, a chess player of unusual skill, who might have become a Master if he had had sufficient time. He lived only a few years after his official retirement; while in London, on 13 March 1933, he died with tragic suddenness. He was a great man as well as a great astronomer. It is said that he had many friends, few critics and no enemies.

On his retirement Innes was succeeded by his Chief Assistant, H. E. Wood, who had come to the Transvaal Observatory in 1906, and who had used the Franklin-Adams telescope in 1910 to take what were probably the best photographs of Halley's Comet secured at that return. (We must wait until 1986 to obtain better ones; this is when the comet is next due.) The best comment on Wood's régime is that it was a time of consolidation. The 26½-inch was now in full working order; altogether over 3 000 new doubles were found with it. Moreover, publications were regular. Wood's own speciality was the discovery and observation of asteroids. He found many, and, as discoverer, was entitled to name them; it is not surprising that South African names were much to the fore, and today's catalogue of minor planets includes for instance, Pretoria, Impala, Knysna, Rhodesia, Outeniqua, Umtata, Pongola and Nyanga. Up to 1938 the Union Observatory had been responsible for the detection of 579 new minor planets – easily a record for any institution. Wood himself had found the first, in 1911, and had named it Transvaalia. Other celebrated asteroid-hunters at the Observatory have been C. Jackson, E. L. Johnson, J. A. Bruwer and H. van Gent.

It was in 1938 that the liaison with Leiden was finally cemented. The Union Government provided funds for the setting-up of a new observatory in the main grounds, and in it was installed the 16-inch twin Rockefeller telescope. It remained there until 1957, when it followed the Franklin-Adams telescope to the official outstation at Hartebeespoort.

In every way Wood was a splendid Director, and he too was a man of many interests (he even published an official paper about South African earthquakes). When he retired, in 1941, his obvious successor was Willem Hendrik van den Bos, who had come to Johannesburg as a temporary visitor so long before and had never returned home.

Van den Bos was an ideal choice, partly because of his long association with the Observatory, partly because of his link with Leiden, and perhaps mainly because his interests

The Republic Observatory, Johannesburg, viewed from the Old Meteorological station.
– Dr Patrick Moore

were very much in the Johannesburg tradition. He had been born in Rotterdam in 1896, and, significantly, his doctorate thesis was on the subject of double stars. Like Innes, he was primarily a double-star observer, and for him the 26½-inch was a perfect telescope. He made a total of 71 929 measurements, discovering 2 895 new pairs – a record which is not likely to be surpassed – and compiling a catalogue of double stars south of declination –19° which remains the standard. There were few cloudless nights when he was not to be found in the dome, hard at work. Quite apart from his actual observing, he worked out the orbits of nearly one hundred binaries, using methods of his own invention which promptly came into general use. Naturally enough he came to be accepted as the leader in his field, and he received many honours – notably the Gold Medal of the Royal Danish Academy, and the Gill Medal of the Astronomical Society of Southern Africa, of which he was President in 1943 and again in 1955. He also found time to discover a hundred new minor planets. It was an amazing achievement.

During his Directorship, the Observatory improved its time service; new, modern-type quartz clocks were installed, and it was said that Johannesburg became "the Greenwich of South Africa". The Observatory's reputation was even more enhanced, and the greatest credit was due to van den Bos himself, who, incidentally, was a man with a charming personality and absolute integrity. After his retirement, in 1956, he continued to observe as energetically as ever until ill-health finally forced him to give up ten years later. He died on 30 March 1974. By then, the clouds over the Observatory – now, after the withdrawal of South Africa from the Commonwealth, re-named the Republic Observatory – had begun to gather.

Van den Bos had been succeeded by Dr W. S. Finsen, who had devoted his life to the Observatory and who was yet another outstanding astronomer who specialized in double stars and in minor planet work. Finsen's achievements matched those of any of his predecessors, and he was responsible for some vital improvements in instrumentation; he it was who developed what is called the eyepiece interferometer, a device which could be attached to a telescope to help in measuring very close double stars. Finsen found 73 new pairs, examined over 8 000 stars, and measured the orbital motions of many binaries. For this work, of course he used the 26½-inch, generally with a magnification of 620. He also concentrated upon photography of the planet Mars, and obtained the best results available before the first space-probe flew past the Red Planet in 1965. Finsen took 54 000 photographs of Mars altogether. There was yet another branch of research in which he was pre-eminent; his observations and predictions of the first artificial satellites made up a vital link in the chain.

It was true that the growth of Johannesburg meant that there was far more sky-light than there had previously been. The Franklin-Adams telescope was accordingly moved to the Hartebeespoort outstation in 1954, which was a good idea; the essential factor for its maximum efficiency was sky darkness – though this did not apply to the great 26½-inch refractor, with its very long focal length. At Hartebeespoort, conditions were still excellent, and there seemed no reason to suppose that they would deteriorate in the immediate future. It was desirable to augment the equipment with a large new reflector, and in 1964 Finsen and van den Bos proposed that a 74-inch should be erected at Hartebeespoort. The mirror blank was actually obtained in England. This was also the year in which the Observatory was officially transferred to the CSIR (Council for Scientific and Industrial Research). The outlook seemed bright – but it was not, so far as Johannesburg was concerned.

The first official rumblings came on 23 September 1970, when the CSIR authorities announced that there had been agreement with the British Science Research Council for the setting-up of the combined South African Observatories; the Cape would be its headquarters, and Sutherland the outstation, 235 miles away across the Karoo, with Sir Richard Woolley as Director. This meant – as we have already noted – that the main Cape telescopes would be moved to Sutherland, while the Radcliffe Observatory at Pretoria would be dismantled altogether. But what about Johannesburg?

There were sound reasons for transferring the Cape telescopes. Seeing there has never been of the best. Reluctantly, it had to be admitted that the same had become true of Pretoria; the 74-inch could not function to full effect under the new conditions – and Pretoria was still growing steadily year by year. But as Finsen, van den Bos and others were quick to point out, the situation with regard to Johannesburg was not the same. The 26½-inch refractor was not badly affected by the artificial lights, at least in the programmes for which it was being used (and had been used ever since its completion). Moving it would be a truly Herculean task in any case. In an article in *Nature*, dated 29 January 1971, van den Bos wrote as follows:

"The 26½-inch refractor, with which the Observatory built up its reputation in double-star observation, is little, if at all, affected by sky illumination or smog, and the time-tested 'seeing' remains as good as ever. There is no reason why it should not be usefully employed at this work, at its present site, for many years to come . . . Even if the observatory were moved *en bloc* to Sutherland it would remain more than doubtful whether its largely astrometric programmes could be continued there as successfully as in Johannesburg or Hartebeespoort . . . The minor planet work would be seriously affected by transfer from the clear winter skies of the High Veld, of that there is no doubt, for it is in winter, when the ecliptic is south of the equator, that the programme reaches its peak. Furthermore, double-star work in particular is of an intensive, long-term and personal nature."

Finsen was in full agreement that to close the Republic Observatory would be a tragic miscalculation – and he was not alone in his views. Yet in many ways the situation was curious; after Finsen's official retirement as Director in 1965 no successor had been appointed, and although the programmes were being continued much of the impetus had gone. This was despite the addition, in 1968, of a new 20-inch telescope bought from America at a cost of 35 000 rands, which was electronically controlled and which was proving to be of tremendous value.

Other arguments also were advanced. Killing the Republic Observatory would leave South Africa's largest city without any major astronomical institute, and presumably the Library, second only to that at the Cape, would also go. Shifting the 26½-inch itself would have been impracticable even if it had fitted into the proposed programme for Sutherland – which it did not. The situation was still fluid when the International Astronomical Union, the controlling body of world astronomy, met in 1972 (it meets every three years; the 1972 venue was at Brighton, in England). Misgivings were expressed by the Commissions dealing with minor planets and with double stars. In the official report published after the congress there was a comment from Dr P. Herget, the famous American asteroid observer, to Dr F. J. Hewitt, Vice-President of the CSIR in Pretoria: "There have been more observations of more minor planets made at the Johannesburg Observatory (and the annexe at Hartebeespoort) than at any other observatory in the Southern Hemisphere in the whole history of astronomy. To destroy this treasure-trove of observations will surely bring you lasting and increasing condemnation as the years go on." And from the Double Star Commission, its President, Dr J. Dommanget, wrote that "unfortunately, as a consequence of recent re-organizations of the astronomical research in South Africa, practically all research work in the double star field in the southern hemisphere has been interrupted. It is imperative that immediate interest is paid by some astronomers in double star research, and effective support given to continuous observations".

It is probably true to say that the proposed demise of the Republic Observatory caused arguments which were more heated than any in the whole story of South African astronomy. At the Observatory itself some work continued. George Frederick Knipe, born in England in 1916 and a member of the Union Observatory staff since 1958, continued to use the 20-inch telescope for photoelectric photometry; he gave up only when the telescope was dismantled in 1972 and sent to Sutherland. (Knipe himself died in the following year.) In 1973 extensive photography of Mars was carried out with the 26½-inch, mainly by Gregory Roberts, in conjunction with the "Mars Patrol" organized from the Lowell Observatory in Arizona; the telescope was then working excellently, and one of the present writers made some observations of Mars with it. The last professional astronomer to work at the Republic Observatory was J. A. Bruwer, who strove, virtually staffless and alone, to complete his personal research in connection with minor planets. But by the end of 1975 only Bruwer was left of the official astronomers.

To leave the 26½-inch idle would be deplorable. Full use of it was offered to the amateurs of the Astronomical Society of Southern Africa, who, however, did not feel able to take on the responsibility of so large an instrument. At the moment (1976) the 26½-inch is fully functional, and there is no thought of shifting it, so that its story is not over even though at the moment it is lying fallow. Meantime, the amateurs have their own dome in the Observatory grounds, containing a 12-inch reflector.

One suggestion put forward was to set up an administrative headquarters at the old site, leaving the 26½-inch there for continued use on its own special programmes and with an outstation further away from Johannesburg; but it was not to be. What will happen in the future only time will tell – but at least the Republic Observatory has an honoured place in history. It, and its great observers such as Innes, Wood, van den Bos and Finsen, will always be remembered as having laid the foundations of modern-type work on double stars and asteroids.

6. THE LEIDEN STATION

HOLLAND is one of the smaller European nations, but it has always been in the forefront of astronomical research. Dutch names appear over and over again – right back to the time of Christiaan Huygens in the seventeenth century, who was probably the best observer of his time and who discovered the true nature of Saturn's ring-system. In modern times Dutch radio astronomy has been of vital importance. Then there was Willem de Sitter, a colleague of Einstein, who was one of the pioneers of relativity theory – and it was de Sitter who provided the first direct link with South African astronomy, because he was a close friend of Robert Innes.

Innes, as we have seen, was an internationalist, and he was always anxious to collaborate with astronomers in other parts of the world. In 1923, mainly because of his association with de Sitter, he came to an agreement with Leiden University, which has always had an exceptionally strong Department of Astronomy. Basically, it was arranged that Leiden astronomers could come to Johannesburg any time they liked, and would have full access to the equipment there; conversely, astronomers of the Union Observatory would have similar facilities at Leiden. In practice the traffic was principally one-way, because there can be no doubt that as an observing site the Transvaal is infinitely superior to the Dutch lowlands. Many Leiden astronomers came on visits, and some of these visits were protracted; that of Willem van den Bos proved to be permanent.

In 1938, three years before van den Bos became Director and during the final part of Wood's régime, the liaison with Leiden was cemented. The South African Government provided funds for the setting-up of a new building in the grounds of the Union Observatory, to become an official station of Leiden University; it was equipped with a 16-inch twin astrograph, known as the Rockefeller telescope (not to be confused with the later Rockefeller telescope, the Boyden 60-inch). The Leiden instrument was made up of two telescopes, each with an object-glass 16 inches across, to be used for stellar photography. The telescopes were identical, and the arrangement was designed as a safety-measure. When a portion of the sky was to be photographed, both telescopes were loaded, so that if one plate proved to be faulty for any reason the other would probably be all right.

Everything proceeded according to plan. The flow of Leiden observers continued, except (naturally) during the war years. But then came a new development. Though the spread of Johannesburg did not materially affect the long-focus 26½-inch refractor, it did reduce the effectiveness of the other telescopes, and a new site was clearly desirable. It was set up at Hartebeespoort, in idyllic surroundings close to the dam, where the seeing conditions were excellent and the darkness of the sky unimpaired. The first telescope to go there was the Franklin-Adams instrument, which had been used for much of the original Franklin-Adams sky chart but which had become rather outdated. The 16-inch twin followed three years later, in 1957, just after Finsen had taken over the Directorship from van den Bos. To Leiden went Dr Theodore Walraven, one of the world's most celebrated astronomers. The plan was to collect observations from Hartebeespoort, with its pure skies, and send the results back to Leiden for analysis. The choice of site was certainly a good one. (It has been established that there was a Stone Age colony in the same place over a thousand years ago, so that ancient men also must have liked it.)

The Rockefeller 16-inch twin astrograph at Leiden Southern Station, Hartebeespoort.
– *David Stevenson*

The arrangement at that time was clear enough. The station at Hartebeespoort was operated by Leiden observers, but it was an official outstation of the Union Observatory, and as such came under Finsen's jurisdiction even though it was virtually autonomous. This was all quite satisfactory, and in 1957 there came a new, powerful instrument: a 36-inch reflector, which was set up in an elaborate observatory with a run-off roof. With it, Walraven and his colleagues worked untiringly – and at the time of writing (1976) they still do.

The 36-inch, known generally as the flux collector, has various unusual features. It is entirely automatic in setting. If the observer wants to turn the telescope to any particular point in the sky, all he has to do is to retire to the control panel at one side of the observatory and push the relevant buttons; the telescope then swings majestically to the position needed. The setting in this way is accurate to within 10 seconds of arc, which is true precision. The optical system follows the Cassegrain design, but the telescope is not intended for visual use. Its rôle is to study the stars in the light of various colours; the starlight is split into five channels, and these are fed into devices known as photomulti-

Dr Theo Walraven demonstrating the observing position for the Rockefeller 16-inch twin astrograph.

– Dr Patrick Moore

pliers, set up in the position which, with a visual telescope, would be occupied by the eyepiece. The starlight reaching the photomultipliers is converted into electrical energy, which is then amplified and sent through a voltmeter. Finally, the results are recorded on punch tape.

Obviously this is research of a highly specialized type, and the 36-inch is as up to date as any telescope in the world. The electronic specialists at the observatory, headed by David Stevenson, take great care to ensure that it continues to perform faultlessly. Nothing could be further removed from the old style telescopes. It is also worth noting that the Hartebeespoort programme never overlapped that being undertaken in Johannesburg itself. This is not to say that the one branch was more important than the other; they ran in parallel.

As recently as 1964 Finsen and van den Bos were advocating Hartebeespoort as a good site for a 74-inch reflector, but, as we have seen, this idea never advanced beyond obtaining the mirror blank. Then came the decision to close the Republic Observatory,

The 36-inch Flux-collector telescope at Leiden Southern Station, Hartebeespoort.
– *Pete Collins*

which was duly carried out. The position at Hartebeespoort was uncertain. Not that there was any suggestion of closing it also; this would have been quite illogical, because the work being carried out there was different from any other, and would not have fitted into the general pattern of Sutherland. Yet Hartebeespoort was officially, if not in practice, still an outstation of Johannesburg.

There, for the moment, the matter rests. At Hartebeespoort, Walraven and his team continue their important work – quietly, without fuss, and with the maximum of efficiency. There is every reason to hope that they, and their successors, will continue to do so in the future.

7. THE YALE OBSERVATORY, JOHANNESBURG

YET ANOTHER South African observatory which has come to the end of its career is the Yale, whose building is still to be seen close to the Planetarium at the University of the Witwatersrand. It too had a distinguished history, but it ceased operations in 1952 simply because the main programme had been completed – and, moreover, conditions so close to Johannesburg had become frankly intolerable.

The main work of the Observatory was concerned with the distances of the stars. The method used was that of trigonometrical parallax, which has already been described. Alpha Centauri, Henderson's original target star, lies at 4,3 light-years, and with the exception of Proxima, discovered by Innes at the Republic (then the Union) Observatory in 1912, is the nearest known star to the Sun; neither is it likely that there are any stars still closer – they would surely have been found by now.

Parallax shifts are very small. Even that of Proxima is less than a second of arc. The early observers had to depend upon direct measurements made at the eye-end of a telescope, but later on photography took over. The procedure was to photograph the same area of the sky twice, with an interval of six months between the exposures. In the interim, the Earth would have moved from one side of its orbit to the other, and any close star would show an appreciable parallax shift. The blink-microscope, to which Innes paid such attention, helped immeasurably in the plate comparisons.

The trouble is that few stars are as close as 50 light-years, and with increasing distance the parallax shifts become smaller and smaller. Beyond a few hundred light-years, the shifts become so tiny that they are swamped by unavoidable errors of observation, and so the method is very limited. Yet it provides the essential basis for all our distance-gauging, beyond the Solar System, and with the nearby stars it is very accurate indeed. Further out, astronomers have recourse to less direct methods, which are bound to involve greater uncertainties.

One of the pioneers in trigonometrical parallax work was an American astronomer, Frank Schlesinger, who became Director of the Yale Observatory. He it was who initiated a major programme, ending in the production of what are known as the Yale Zone catalogues. He measured the parallaxes of many stars, together with their proper motions; but from his own country he could not reach the stars of the far south – and this was where Johannesburg came into the story.

What Schlesinger wanted to do was to set up a special observatory, to extend the parallax survey down to the south celestial pole. For this kind of research a large refractor was ideal, but the main South African telescopes, such as the 26½-inch, were already fully engaged on programmes of their own. So, with full permission and co-operation, the Yale authorities decided to proceed independently. In Braamfontein, outside Johannesburg, on the campus of the University of the Witwatersrand, they set up an observatory of a very unusual kind. There was no majestic dome, or even a turret; the observatory took the form of an asymmetrical brick structure, with a slit in the roof which could be opened but which did not rotate. To measure a star's parallax, it is essential for the star to be high in the sky when the relevant photographs are taken – so there was no need for the telescope to be wholly manoeuvrable, or to cover the whole sky. So long as it could be pointed some way to

The Old Yale Observatory building.

– Dr Patrick Moore

either side of the meridian, it would serve its purpose very adequately. Needless to say, a building of this sort was far cheaper than a dome.*

The telescope provided was a 26-inch refractor of long focal length – very similar in aperture to the Union telescope. By 1925 all was ready, and the observer, Dr. H. D. Alden, began his work. It was routine, but none the less vital for that. Alden made thousands of measurements, took almost innumerable photographs, and made a major contribution to the Yale Catalogue. He also studied variable stars, asteroids, and the satellites of the principal planets, but these researches were always subordinate to the main task. Alden

*There was one great telescope which was even more restricted, and whose tube was swung between two massive stone walls, so that it could point only a short way to either side of the meridian. This was the 72-inch reflector built at Birr Castle, in Ireland, by the third Earl of Rosse in 1845 – and with which he discovered the spiral forms of the systems we now know to be external galaxies. However, Lord Rosse followed this pattern not because he wanted to, but because with the limited engineering facilities available at the time he had no choice. The 72-inch was last used in 1909, and is still to be seen in the Castle grounds. Up to the middle of the first world war it remained the largest reflector in the world, though it had a metal mirror, and was by no means the equal of some of the smaller but newer instruments.

retired in 1945; he had accomplished much (he also served as President of the Astronomical Society of Southern Africa in 1931–2). He was succeeded by an equally energetic observer, C. Jackson. The only change in the organization after the end of the war was that from 1 January 1946 the station was operated jointly by the Universities of Yale and Columbia.

As time went by, the programme approached completion. Even the 26-inch could reach only a limited number of stars, and after Jackson had been observing for some years it began to seem that he had done all that was really practicable. The telescope could not be turned over to more general work, because of the restrictions imposed by its mounting. Moreover there seemed little sense in persevering on the old site, because the growth of Johannesburg had caused a marked deterioration in the seeing conditions which seemed likely to accelerate. Even if the telescope had been as far afield as the Union Observatory, there would have been some basis for discussion. As things were, the decision was made to close the observatory and transfer the equipment to Canberra in Australia. Jackson remained at his post until the final days, but in 1952 the Yale Observatory was finally dismantled. Today only the building remains, and there are no astronomical instruments inside it.

The story of the Yale Observatory is exceptional. It was set up for a specific purpose, and was never intended to function indefinitely; it accomplished its task, and passed gracefully into history. Its work had been well done.

8. THE RADCLIFFE OBSERVATORY: END OF A STORY

NO UNIVERSITIES in the world are more celebrated than those of Oxford and Cambridge. Today Cambridge is particularly noted for its astronomical research, both optical and in the radio range; the present Astronomer Royal, Sir Martin Ryle, is based there instead of at Greenwich. Oxford, too, has its observatory, but the story is rather different, and much more closely linked with South Africa.

The observatory at Oxford was founded in 1772 by the Radcliffe Trustees, and was therefore always known as the Radcliffe Observatory. Telescopes were set up, and were extensively used; but nothing can be done about the British climate, and so far as optical viewing is concerned there are serious limitations. Moreover, even in the early 1900s Oxford was spreading; from being a university with a small town tacked on, it was turning rapidly into an industrial city, with all the benefits of urban civilization – such as smoke, grime, fumes and smog. Clearly it was desirable to move elsewhere, and the man who seems to have provided the first impetus was the then Radcliffe Observer, Harold Knox-Shaw, who had taken up his Oxford duties on 18 August 1924.

Knox-Shaw was born at St Leonards on Sea, Sussex, on 12 October 1885, the eldest of a family of four. He was educated at Wellington College and then at Trinity College, Cambridge, where he graduated Sixth Wrangler in 1907. While at Wellington he came across the mathematics master, S. A. Saunder, who was a famous amateur observer of the Moon; and Knox-Shaw's interest in astronomy dated from that time.

After graduating he went to Helwan in Egypt, to be assistant at the Khedivial Observatory. With the 30-inch reflector there, he photographed southern nebulæ, and also took the first photograph of Halley's Comet at its last return; he recorded it on 24 August 1909 – the first man to have detected it since Sir John Herschel's final observation from the Cape. Knox-Shaw became Superintendent at Helwan in 1913, and left only in 1924 to take up his post at Oxford, in succession to A. A. Rambaut, who had died ten months earlier. (During the interim the authorities had called in J. L. E. Dreyer, former Director of the Armagh Observatory, who had been living in retirement in Oxford.)

Knox-Shaw, used to the clear Egyptian skies, was not in the least impressed by observational conditions at Oxford, and he soon began laying plans for a move. He had the support of Sir Frank Dyson, the Astronomer Royal, and the change soon came. In 1929 the Radcliffe Informary felt the need for expansion, and offered to buy the nine-acre site on which the Observatory had operated for the previous hundred and fifty years. By then Knox-Shaw and Dyson had already made a preliminary reconnaissance to South Africa, and on their return it was tacitly agreed that the shift should be made. In November 1929 an agreement of sale to the Medical School was announced.

In only remained to find a suitable location in South Africa. Pretoria had a good climate, and was convenient in many other ways as well, but it was essential to be quite sure. The Trustees decided that the man to find out was Dr W. H. Steavenson, who has already been referred to in this book. Steavenson came to Pretoria, and carried out an extensive series of tests from a hilltop to the south-east of the city. Steavenson's report was favourable. The Municipal authorities then stepped in, and offered to make the Radcliffe Trust a gift of the 57-acre site; they even promised to lay on water and electricity.

It was a generous offer; the Pretoria authorities kept their word, and continued to give all possible help throughout the Observatory's career.

Unfortunately, an Oxford faction led (typically!) by Professor F. A. Lindemann, afterwards Lord Cherwell, caused prolonged delays by opposing the transfer of Trust capital to an area outside the jurisdiction of the English courts. Among those who were called in to support the move were Knox-Shaw, Dyson, Sir Arthur Eddington, Schlesinger, Shapley, de Sitter and Spencer Jones. Eventually, in July 1934, the battle was won, and in August 1935 it was possible to sign a contract with the famous firm of Grubb Parsons, in Newcastle, for the construction of a 74-inch reflector and a 61-ft. cylindrical turret to house it. The 74-inch would be the largest telescope in South Africa (in fact, it still is) and the outlook seemed bright.

Knox-Shaw himself was already in residence at the new Observatory by 1939, with R. O. Redman as Chief Assistant and E. Gwyn Williams as Second Assistant. The Observatory buildings were ready, even to the turret for the 74-inch which Knox-Shaw had strongly urged in lieu of the more conventional dome. Unfortunately the telescope optics were not only not ready, but were not even fully under way. The pyrex disk for the 74-inch had been ordered from the Corning Glass Company in New York, probably the world's leading manufacturers of these expensive items. The first two disks had to be rejected. The third was successful, but it did not reach Newcastle until October 1938; and before it had been ground, polished and tested, the War broke out. It is not surprising that the making of the mirror took a long time – it weighs 36 hundredweight, and every part of the surface must be correct to within a millionth of an inch – and of course, all work on it had to be suspended until after the German defeat. Only in 1945 was it possible to return to civilized pursuits, and the mirror did not reach Pretoria until April 1948, nearly thirteen years after the original order for the disk had been cabled to the Corning Glass Company.

On the arrival of the mirror, Knox-Shaw's staff had been reduced to one assistant (David S. Evans) and three African labourers. Nevertheless, it took very little time to get observing programmes started. Knox-Shaw retired in 1951, so that he had the benefit of the 74-inch for only two years; but he had twenty years of happy retirement at Elgin, Cape, watching the triumphant progress of the Observatory to which he had devoted so much of his life. He died in Cape Town on 8 April 1970; his wife and their one son survived him.

Knox-Shaw was a man of great charm and sense of humour as well as absolute dedication. There is a story that on one occasion he had a bet with Eddington about the value of a certain mathematical constant. As it was unlikely that the matter would be cleared up during their lifetimes, it was agreed that the loser should ultimately present the winner with a new harp. Most astronomers agree that by now the bet has been fairly and squarely won by Knox-Shaw . . .

The Observatory lay on the range of hills to the east of the old Boer fort of Klapperkop, five miles from the centre of Pretoria and at an altitude of 5 000 feet. The southern aspect was clear and lightless as far as Johannesburg – a state of affairs which, alas, did not last; before long brilliantly-lit buildings began to make themselves unpleasantly obtrusive. The road from Pretoria was long, attractive and wooded. It was also something of a residential area, and, appropriately, many of the roads were re-named according to a cosmical

"Canopsus" Street.

– *Ludolf Meyer*

pattern: Orion Street, Perseus Street and so on. One of these new names gave rise to an episode which still tends to make astronomers chuckle.

The second brightest star in the entire sky is Canopus, in the constellation of Carina, the Keel of the old Ship Argo. Naturally, it had to have a road named after it, but the local authorities accidentally inserted an extra S, and up went the official sign: *Canopsus* Street. Canopsus Street it remained, but after many years a distinguished astronomer, then resident in Pretoria, wrote to the authorities and pointed out the error. The reply was stunning. "The name *Canopsus* is on all our maps, and has been officially accepted," they wrote. "It cannot be altered now. Can't you change the name of the star?"

It is not very likely that such an idea appealed to members of the International Astronomical Union. Ironically, the road has now been renamed *Canopus*, but the Radcliffe Observatory no longer exists . . .

But this is running ahead of the story. Way back in 1948 the Observatory buildings were stately and imposing, and the grounds were gradually put into order, so that eventually less than a dozen poisonous snakes were, on average, killed each year. Accommodation for the resident astronomers and their families was built, and before long the Observatory

The Radcliffe Observatory, Pretoria, in 1952.
– *Courtesy of Africana Museum, Johannesburg*

became something in the nature of a small, specialized township. But the heart of it was the 74-inch, in its unusually-shaped turret. The non-rotating parts of the building were made of brick; to ensure that the temperature inside remained relatively constant, the brick wall contained a cavity which was surrounded at a distance of six inches by an outer sheath of metal lathing, coated with rough-cast plaster. The revolving turret, too, was double-walled; the outer cover was made of galvanized steel and the inner of wood-fibre board. Great care was taken in temperature control, and it was possible to keep the temperature steady within a range of 5° Fahrenheit at the level of the mirror, whereas outdoors the diurnal range was the order of 25° Fahrenheit.

The 74-inch was conventional enough. The tube was made in three parts; the cell containing the mirror (total weight almost five tons), the fabricated steel centre section connecting the tube to the declination axis, and the skeleton or lattice upper section, made of aluminium alloy, carrying the secondary mirror. The telescope could be used either at the Newtonian focus, or as a Cassegrain or Coudé for spectroscopic work, which occupied the major part of the observing time during the period when the 74-inch was operating from Pretoria. There were one or two special features; for instance, the planners had little faith in the constancy of the local electricity supply, and the drive mechanism of the telescope was actually regulated by a tuning-fork!

Once it had been completed, the 74-inch was used on every clear night. Special programmes included studies of the radial velocities of stars; the structure of the Magellanic Clouds; investigations of globular clusters, and so on. A very important part of the Observatory's life was its international aspect. Astronomers from all over the world came there, to do specialized research and also, in many cases, to complete their degrees (one such visitor was Brian Warner now Professor of Astronomy at the University of Cape Town, whose PhD thesis concerned late-type stars). There were strong ties with the Science Research Council, in London, as well as with Oxford.

On Knox-Shaw's retirement, in 1951, Dr A. D. Thackeray became Director. It was under him that the Observatory enjoyed its most productive years. The 74-inch did all that could have been expected of it; so far as South Africa was concerned it reigned supreme, and for many years all went well.

Thackeray, a Cambridge graduate, had carried out post-graduate work at Mount Wilson, in California, before returning to Cambridge. He came to South Africa in 1948, just in time to help with the final testing of the 74-inch, whose mirror had just arrived after what must have seemed an interminable delay. In the following years streams of publications came from Radcliffe, not only by the resident staff but also by distinguished visitors – such as Professor W. N. Stibbs, now Director of the University Observatory of St Andrews, in Scotland, who spent many months at Radcliffe studying variable stars. Thackeray's own special work was in connection with stellar spectroscopy, and here, of course, the tremendous light-grasp of the 74-inch was invaluable.

Radcliffe was now world-famous, and it was a friendly and welcoming place. The amateurs of the Astronomical Society of Southern Africa erected a dome containing a 12-inch reflector, almost in the shadow of the great turret of the 74-inch, and were always given every possible encouragement.

It seemed too good to last – and it was. Pretoria was spreading its tentacles over the countryside; and even if Canopsus Road was not engulfed, the general increase in sky illumination became alarming. Normally the astronomers would have carried on, doing

The Radcliffe Observatory, Pretoria, in 1975.
– *Dr Patrick Moore*

no more than mutter imprecations against the builders of the neon lights; but on 1 January 1972 came the announcement of the merger of the main South African observatories and the setting-up of the new centre in Sutherland. For Radcliffe, the writing was very clearly on the wall.

It must have been a strange period. The Observatory was a dying institution from the moment that the decision was made to shift the 74-inch to Sutherland – and yet the work went on as before. Thackeray retired officially as Director in April 1974, but he and his remaining staff went on making the most of every clear night. With him, to the very end, were Drs M. W. Feast,* Peter Andrews and R. Catchpole, busily engaged in using the reflector to carry out photometry of seventeenth-magnitude stars (and stars as dim as this are difficult objects indeed). It was only when the telescope was physically dismantled that the work stopped, the observers and their families departed, and the Radcliffe Observatory ceased to exist. Even the turret was taken down, and when the present writers visited the site, in 1975, it looked broken and ghostly. Weeds covered the once-trim paths and gardens, the remnants of the dome were in a sorry state, and the whole atmosphere was one of desolation. By now the final knell has been rung; the amateur dome too has been taken down and erected in the grounds of a school miles away.

Obviously it was a pity that Radcliffe, with its great history and its happy atmosphere, had to go; but one has to admit that from the scientific viewpoint the decision was logical. Now that the 74-inch has been re-erected at Sutherland, it can perform far better than it would have been able to do in the outskirts of modern Pretoria. Sentiment and science do not, usually, mix; but there will be many astronomers who will have nostalgic recollections of those days between 1950 and 1974, when so much vital work was carried out. During its quarter-century of active life, the Radcliffe Observatory accomplished more than many other observatories have done in hundreds of years.

*Dr M. W. Feast was appointed Director of the South African Astronomical Observatory on the retirement of Sir Richard Woolley on 31st December, 1976.

9. BOYDEN: THE OBSERVATORY ON THE HILL

Bloemfontein, in the Orange Free State, is one of the smaller of South Africa's principal cities. Its history has been turbulent at times; today it is both peaceful and pleasant, and it is important astronomically because of its near-perfect climate from the observer's point of view. We have already seen how the setting-up of the great observatory at Sutherland meant the end of the story for some past stations, notably the Radcliffe at Pretoria. The same arguments could not apply to the Boyden Observatory, at Bloemfontein, because in climate it stands up well by comparison with Sutherland or anywhere else.

Boyden has, indeed, one major advantage over Sutherland: it is more accessible, and there is no need for a long drive across virtually uninhabited Karoo. An hour's flight from Jan Smuts in Johannesburg to Bloemfontein Airport, and an uneventful drive along fifteen miles of road, brings you to the wooded hill atop of which the Observatory stands; the setting is picturesque, and there is a pleasing lack of light pollution.

As a matter of fact, the South African phase of the Observatory does not date back very far, and in point of time it is junior to the Cape by more than a century. The name comes from a gentleman named Uriah A. Boyden, who was a mechanical engineer in the American city of Boston. In 1879 he left a sum of 238 000 dollars to Harvard College for the express purpose of furthering astronomical research, and in 1887 the Trustees of the Fund transferred it lock, stock and barrel to the Harvard College Observatory, which was then – as now – one of the major establishments in the United States.

One of the main branches of work at Harvard has always been stellar spectroscopy. The original work on classifying stellar spectra was undertaken by men such as Angelo Secchi in Italy and Sir William Huggins in England, but their preliminary systems were soon superseded by that of Harvard. Each stellar type was given a letter of the alphabet, and the series ended up with the well-known sequence; O,B,A,F,G,K,M,R,N and S – O stars having the hottest surface, and R, N, and S the coolest. Initially it was thought that this series was indicative of stellar evolution; a star would end its career as a dim Red Dwarf (type M or later), while an "early" type spectrum of, say, B or A would denote a star at the peak of its output. Actually, things are much more complicated than this; but toward the end of the last century data were still being collected for analysis, and Harvard astronomers such as E. C. Pickering and Miss Annie Cannon were anxious to extend their surveys as far as possible. This meant that a southern station was if not essential, at least highly desirable. This was where the Boyden bequest came in.

Where should a new station be put? The obvious choice was South America, where conditions are excellent, and where there are lofty sites well above the worst of the Earth's irritatingly dirty and unsteady atmosphere. Surveys were made, and the choice fell upon Arequipa in Peru. The southern Harvard station was duly set up, and it operated reasonably well; certainly much work was done there. And yet it was not ideal. The meteorological conditions did not really come up to expectations, and there were communications difficulties as well; even today Peru is not one of the most accessible of places. After some years had passed, the thoughts of the Harvard astronomers began to turn elsewhere.

Sir David Gill was particularly vocal in stressing the advantages of South Africa, and

An aerial view of Boyden Observatory. The 60-inch Rockefeller Telescope building is seen in the right foreground.

– *M. J. Bester*

Harvard became keenly interested. In 1908 Solon I. Bailey, an eminent American astronomer, arrived to investigate – and he was very thorough. Armed with a portable telescope, he went from place to place studying the skies as carefully as he could. Of course he went to the Cape; he went to many other sites as well, but finally he came down in favour of Bloemfontein. He returned to Harvard full of enthusiasm, only to find that there were major obstacles ahead.

A great deal of equipment had been set up at Arequipa, and moving it from one continent to another would be an expensive operation. So for a long time nothing could be done, and Arequipa continued to function as effectively as it could. The next developments did not come until 1926, by which time the Harvard directorship had been assumed by Dr Harlow Shapley – the man who first measured the size of our star-system or Galaxy. (He was able to do this because of the observations made at Arequipa, as we will see.)

Shapley was far-sighted, and he had immense drive. He was able to obtain generous grants from both Harvard University and the International Education Board, and he took the plunge: Arequipa would be closed, and Bloemfontein – or, rather, Mazelspoort, fifteen miles from the city – would replace it.

In 1923 Dr John Stefanos Paraskevopoulos had become astronomer in charge of Arequipa. He was Greek, born on 20 June 1889; he was a graduate of Athens University, a corresponding member of the Greek Academy, and also a Knight of the Order of Phoenix – as well as being a noted astrophysicist. In July 1927 he came to Mazelspoort together with his wife, who was also an astronomer. They made a final check, and were as satisfied as Solon Bailey had been a decade earlier. The die was cast.

Almost at once the first equipment was shipped over, and the first observations from the new Boyden station were made in September 1927; Paraskevopoulos was officially appointed Director. Yet there was much to be done, and it was only in 1933 that the Observatory was in full working order. The largest telescope was the 60-inch Rockefeller reflector, though its optics were admittedly somewhat antique. There was also a 24-inch reflector known as the Bruce Astrograph because it had been donated to the southern station by a Mrs Bruce; there was the 13-inch refractor made by Alvan Clark, one of the most famous opticians of the last century and who had been responsible for many giant objectives; and there was the 10-inch Metcalf refractor, which from an historical viewpoint was probably the most celebrated of all. Of these telescopes only the Bruce Astrograph has disappeared; in its place was set up the famous ADH telescope, to be described below.

Paraskevopoulus remained Director until 1951; his was a long and highly successful régime. (He died on 15 March 1961.) After his retirement there was no official Director, and it was tacitly agreed that the senior visiting astronomer should be regarded as taking charge. This was by no means a satisfactory arrangement, but it persisted until 1968, when Dr Alan H. Jarrett was appointed Director.*

The 60-inch Rockefeller reflector remains the largest telescope at Boyden, and is the second largest in the Republic; it is exceeded only by the 74-inch formerly at Pretoria and now at Sutherland. Go there, and you cannot fail to be impressed by the dome itself; trees surround it, without doing any damage to the sky-view or the seeing, and the only thing to bear in mind is that Boyden is "snake country"; it is wise to take care when walking through undergrowth. (Snakes, so naturalists say, are more frightened of men than men are of snakes. In the case of the present writers, this would be difficult!)

Seen from the moving platform which holds the main control panel, the telescope, with its latticework tube, seems to reach almost to the top of the building. In fact the length of the "tube" is only about thirty feet, but it is most impressive, and when the platform is raised there is a yawning gap below, through to the base of the massive mounting. The whole telescope weighs 40 tons, and the mirror itself accounts for four tons. Yet the driving mechanism is so precise that very long photographic exposures can be made with purely automatic guiding.

*It is worth listing some of the famous astronomers who were at Boyden for periods during this time: Bart J. Bok, A. G. Velghe, M. Haffner, A. van Hoef, E. M. Lindsay, E. H. Geyer. T. Schmidt, A. D. Andrews, J. Dommanget. For much of this time M. Bester acted as chief observer.

The 60-inch Rockefeller Telescope at Boyden.

– *Dr Patrick Moore*

The name of the telescope has an interesting history. In the early part of the century Harvard obtained two mirrors from a famous English optician, Dr A. A. Common, with the intention of installing them in a southern station. But the cost of a mirror is only a part of the cost of a whole telescope, plus its observatory, and – as so often happens – funds were short. So the mirrors lay unused; and it was only in the 1930s that the Rockefeller family agreed to provide what was needed. It was, then, only fair to call the 60-inch the Rockefeller reflector. But there was another point to be considered: Common's mirrors were good, but not up to modern requirements.

When Jarrett became Director, in 1968, he promptly decided that the telescope would have to have new optics. The cost amounted to about 200 000 rands, but was well worthwhile. The opportunity was also taken to renovate the drive of the telescope, which had become worn over the years, so that when everything was completed the telescope was as up to date as any in the world.

Of course, telescopes of this size are almost never used visually. Until recent years the emphasis has been on photography, but even this is now being superseded by photoelectric devices, which are far more sensitive. Photoelectric photometers, image tubes and the like are used with the Rockefeller reflector, and the large number of clear nights means that programmes can be carried on throughout the year with practically no interruption.

Variable stars have figured largely in the programme, and the Rockefeller telescope is also used for photographic and photoelectric observations of stars and starfields in the southern Milky Way and in the all-important Magellanic Clouds. In addition there is a sky patrol programme which is the counterpart of the Harvard survey of the northern hemisphere. One convenient feature of the 60-inch is that the change-over from one optical system to another can be carried out quickly and smoothly.

Though the 60-inch is pre-eminent, Boyden also contains other instruments. In 1976 the list was as follows:

1. *The Rockefeller reflector*. Aperture 60 inches (152 cm), focal length 798 cm. Used with various pieces of accessory equipment, including a very modern photometer and a Jena grating spectrograph.
2. *The ADH Baker-Schmidt telescope*. Aperture 36 inches (91 cm), focal length 303 cm. With it there is an objective prism, angle 3°10′, aperture 84 cm. The ADH is at present (1976) being renovated, and is out of use for the first time since it began its career in 1950.
3. *The 13-inch refractor*. Aperture 13 inches (33 cm), focal length 482 cm.
4. *The Metcalf Triplet*. Aperture 10 inches (25 cm), focal length 124 cm. An objective prism (angle 2°5′) can be used with it. This is the famous telescope used to take the Magellanic Cloud pictures which will be described below.
5. *The Nishimura Reflector*. Aperture 16 inches (41 cm). Used for flare star research; it has a cooled, solid carbon dioxide photometer for this purpose.
6. *The Hamburg Schmidt*. Aperture 14 inches (36 cm). Not in use at the moment.

There are also various cameras. As we have noted, the list during the earlier part of the Observatory's story was much the same; the only major change has been that in 1950 the ADH telescope replaced the old 24-inch Bruce Astrograph.

Despite its current inactivity, the ADH telescope is second in importance only to the Rockefeller. Here we have an example of true international collaboration, though, alas, politics has managed to intrude during the past few years.

A stands for Armagh, the main observatory in Northern Ireland; D for Dunsink Observatory, outside Dublin in Eire; and H for Harvard. Something must be said here about Armagh, in particular, because it comes very much into the story.

Armagh Observatory – about thirty miles inland from Belfast – was founded in the late 18th century, and during Victorian times an important star catalogue was compiled there by the then Director, a formidable gentleman named Romney Robinson. (Many stories are told about Dr Robinson. It is certainly true that on one occasion he managed to make the Government re-route a major railway line, on the grounds that the vibration of passing trains would prove an intolerable nuisance to the astronomers at Armagh.) Subsequently things became rather difficult, and the Observatory's revival came only in the 1930s with the arrival of Dr Eric Mervyn Lindsay, who had studied astrophysics under Shapley at Harvard. Lindsay had been assistant at Boyden before being called home to his native Ireland to take charge of the ailing Armagh Observatory, where, to be frank, very little research had been done for decades. Lindsay was nothing if not energetic. He revitalized the Observatory, whose main instrument had been a modest 10-inch refractor; he installed an 18-inch Schmidt telescope, able to photograph wide areas of the sky with a single exposure; and he maintained connections with colleagues in many lands. He was a noted theorist as well as an observer; his main research was concerned with the Magellanic Clouds. (He was, incidentally, a close friend of one of the present writers – Moore – who acted as Director of the Armagh Planetarium between 1965 and 1968.)

Eric Lindsay was a man with ideas. His Boyden work with the Magellanic Clouds had shown him the importance of southern stations, and Boyden was an ideal site for a new telescope of the kind needed for this sort of work. He joined forces with the Director of Dunsink Observatory, Professor Brück, and with Shapley, who was still at Harvard. Between them they raised sufficient funds for the construction of a new telescope, and in due course the ADH 36-inch was installed.

The optical system was of the Schmidt-Cassegrain pattern. The main drawback of a more conventional telescope is its small field of view; to photograph the entire sky with, say, the famous Hale 200-inch at Palomar would take an impossibly long time. A new system had been developed in the 1930s by Bernhard Schmidt, an Estonian who had originally been concerned with the manufacture of explosives, but had blown off one of his arms, and had turned to optics as being rather safer. Basically, his idea was to use a spherical mirror rather than a paraboloid, and to correct the inevitable faults by means of a highly complex plate at the top end of the tube. The result is that wide areas can be photographed with good definition right to the edge of the plate, and Schmidts have proved invaluable – in fact, they provide so much information that only recently have instruments been developed for analyzing the results with reasonable speed.

There were no big Schmidts in the southern hemisphere before the completion of the ADH, so that there was endless scope, and for two decades the ADH remained the largest southern telescope of its kind (it is still the largest in South Africa). At once it went into action, and proved to be a complete success. It was used, and used, and used. In fact it was used so much that the drive became worn out, and there was no alternative but to re-make it; it was sound sense to build a new mounting as well. There was the usual problem of finance, and in fact this had not yet been solved. It is to be hoped that the renovations will be completed soon; the ADH is much too valuable to be left idle for longer than is absolutely necessary.

The 13-inch refractor is housed in another building a few yards from the ADH. The objective was made by Alvan Clark, which is sufficient guarantee of its excellence. It was obtained as long ago as 1890, and had been one of the main telescopes in the Arequipa days before being transferred to Boyden in 1926–7. It still functions perfectly, but there is at least one unusual feature about it, as Dr Jarrett pointed out to the present writers while showing them round in 1975.

He pointed to the floorboards. "You'd better be a bit careful," he said casually. "A four-foot cobra lives under there."

We stepped backwards rather apprehensively. "A what?"

"A cobra," said Dr Jarrett, still in the same casual tone. "He lives there, but we don't bother about him; he leaves us alone, and we leave him alone. No problem – live and let live."

Being cowardly by nature, we were not entirely reassured, particularly when hearing the story of how a visiting astronomer, entering the dome ahead of his escort, had found himself face to face with a rearing snake. We kept a wary eye on that floorboard, and were

Professor Alan H. Jarrett, Director of Boyden Observatory, and Patrick Moore inspect the Metcalf Telescope.

– *Pete Collins*

relieved when the cobra did not appear. However, the Clark telescope itself is excellent by any standards. In the past it was used together with a spectrograph in the pioneer work of classifying stellar types; it is outmoded now, because it is too small for the needs of modern research, but it is still used sometimes (despite the cobra), and for its size it is as good as any refractor in the world.

When we migrated to the adjoining observatory, again with a run-off roof, we could not resist asking whether we could expect a rattlesnake or two; but the only occupant was the 10-inch Metcalf refractor, which again came from Arequipa, and was responsible for a fundamental advance in astronomy. This has to do with the Magellanic Clouds and the variable stars known as Cepheids.

Most stars – including the Sun, fortunately for ourselves – shine steadily, but there are some which fluctuate. The first of these variables, Mira in Cetus (the Whale) was identified by a Dutch observer named Phocylides Holwarda as long ago as 1625. Mira has a period of 331 days on average; like many of its kind it is a huge Red Giant star, swelling and shrinking, and changing its output of energy. Other variables complete their light-cycles in a few days. One of these is Delta Cephei, a rather inconspicuous star too far north in the sky to be seen from South Africa. Its behaviour was discovered later in the 17th century by a young deaf-mute astronomer named John Goodricke, whose promising career was tragically ended by his death at the early age of twenty-one. Delta Cephei is always visible with the naked eye, and has a period of 5.3 days; it is regular, so that one always knows how bright it will be at any one moment. Other similar stars were found – for instance Eta Aquilæ, in the Eagle (period 7.2 days) and Kappa Pavonis, in the Peacock (9 days). Thousands of these completely regular, short-period variables are now known. Because Delta Cephei was the first to be discovered, stars of this type are always known as Cepheids.

Cepheids are very powerful stars, so that they can be seen across vast stretches of space. All of them are remote; and of course the classical method of measuring star-distances, trigonometrical parallax, can be used only for our comparatively near neighbours. When Arequipa was set up, astronomers had no real idea of the scale of the universe. Sir William Herschel had suggested that the so-called spiral nebulæ might be external galaxies, but this was the minority view, and Shapley was among those who disagreed with it.

What about the Magellanic Clouds? Superficially they look like broken-off parts of the Milky Way, but they were obviously of great importance, and at Arequipa many photographs of them were taken with the Metcalf 10-inch refractor. The plates were sent back to Harvard for analysis, and a famous woman member of the staff, Miss Henrietta Leavitt, began to study them. In particular she was interested in the Cepheids, and there were plenty of Cepheids to be found.

Miss Leavitt concentrated upon the Cepheids in the Small Cloud of Magellan, and made a surprising discovery. Of course she could measure the periods of the Cepheids by comparing plates taken on different nights; and she found that the Cepheids with longer periods were always brighter than Cepheids with shorter periods. There was a definite relationship, and the significance of it could hardly be over-emphasized, because to all intents and purposes the Cepheids in the Small Cloud were at the same distance from us, just as for most practical purposes it may be said that Johannesburg and Pretoria are the same distance from New York.

Therefore, if the Cloud Cepheids were equally distant, then those which *looked* the brighter must actually *be* the brighter. It followed that the real luminosity of a Cepheid could be found merely by measuring its period of variation. Because the Cepheids in the Cloud are no different from Cepheids anywhere else, it became possible to measure relative distances.

Shapley, at Harvard, was quick to take advantage of this "Period-Luminosity Law". He studied the Cepheids in the great assemblages of stars known as globular clusters, which lie around the edges of our Galaxy; he measured the distances of the Cepheids, and hence the distances of the globulars; and he was able to draw up the first reliable scale-map of the Galaxy. Later, in 1923, Edwin Hubble at Mount Wilson, in California, studied the Cepheids in spiral nebulæ, and was able to prove that these remarkable objects are independent galaxies, far beyond our own. Since then the Cepheids have played a vital rôle in all studies of the distribution and distances of stars and galaxies. Inevitably there have been complications; for instance there was an error in the basic Cepheid scale which did not emerge until the 1950s, and in a famous paper delivered to an enthralled audience in London the late Walter Baade doubled the size of the universe "at a stroke". But everything depended upon the Period-Luminosity Law, and it was the Metcalf telescope which took the essential photographs.

Of course, the work of the telescope was not ended. Since its move to Boyden it has been extensively used, and more than 4 000 plates have been taken with it, mainly for variable star studies. It, too, is in perfect order; modest though it may be in size, its place in scientific history is safe.

Variable stars are of many kinds. There are the long-period stars such as Mira; the short-period Cepheids; the even shorter-period RR Lyræ stars, which are plentiful in globular clusters (note that the two finest globulars in the sky, Omega Centauri and 47 Tucanæ, are in the far south); the irregular variables, and so on. There are also the flare stars, which brighten up abruptly over periods of a few minutes. The first of them was identified in 1948, and over twenty-five are now known. Pioneer work on them has been carried out at Boyden, notably by A. D. Andrews from Armagh, who spent some months at the Observatory; and one telescope – the 16-inch Nishimura – is devoted entirely to them.

All flare stars are feeble Red Dwarfs. One of them UV Ceti, has been known to rise from magnitude 12½ to 5½ in a few minutes, though it soon dies down again. Another example is AD Leonis, in the field with the second-magnitude star Algieba or Gamma Leonis, where the usual magnitude is 9½, and the onset of a flare can make the brightness rise to magnitude 8¾. It is fascinating to see a flare star in action, but a "watch" is tedious, because one never knows when a flare will begin, and there are occasions when many hours can pass by with no detectable activity at all.

Astronomers need to know how a flare progresses; until the observational data have been acquired, no analysis can be attempted. Visual watching is not the answer, if only because it is not reliable enough or precise enough. Very accurate photoelectric measurements are essential, and here too Boyden has taken the lead so far as South Africa is concerned. A special photoelectric photometer attached to the Nishimura reflector is used, and two flare stars, UV Ceti and AD Leonis, are monitored as constantly as possible. Many outbursts have been recorded, and have provided information about the frequencies and light-curves. Moreover, a flare star sends out radio waves during an outburst, and the brightness measurements at Boyden have been correlated with the radio observations carried out at Jodrell Bank and elsewhere. It would be idle to pretend that the cause of the outbursts is yet known, but there may well be an analogy with the much more localized flares seen on the surface of our own Sun. In any event, flare stars are of tremendous importance in modern astrophysics.

Continuity is the essence of flare-star work, and Boyden could hardly be bettered as a site. The Nishimura work is now being augmented by observations made with the Rockefeller 60-inch, and much may be expected from them – if all goes well.

Quite apart from the astrophysical research, comet-hunting has been carried out from Boyden, notably by M. Bester, the long-serving staff observer. Though primarily an astrophysicist, he has also spent many hours in searching for new comets, and he has been very successful. Here again the clear, transparent skies are a great help.

All this sounds promising, but at the moment (July 1976) there are clouds of a different kind over Boyden, and once again politics has reared its ugly head. The Observatory was an international venture, and nations taking part included the two Irelands, England, Holland, Sweden and West Germany as well as the United States. For years the Boyden Council functioned well, but during the 1960s and early 1970s the Communist pressure against South Africa built up. Nation after nation withdrew its support; Shapley died, and in 1972 Lindsay also died suddenly of a heart attack. Finally the Boyden Council disintegrated, and on 30 June 1976 it went out of existence, leaving the Observatory to fend for itself. It was handed over in its entirety to the University of the Orange Free State, in which Jarrett, as Director, was Professor of Astronomy.

Unfortunately the University did not have sufficient funds to maintain a major observatory, and extra help was needed. This may well come from the University of South Africa. In any case it is very much to be hoped that some solution will be found; Boyden is ideally situated, in full working order, and has several invaluable telescopes, including the 60-inch. To dismantle it would be a scientific blunder which could never be fully repaired. If wisdom prevails, Boyden must continue to flourish, and will play as major a part in world astronomy during the coming years as it has done during the past half-century. Overseas visitors will always be welcome; and if they yearn for a little extra excitement – well, there is always the cobra!

10. THE LAMONT-HUSSEY OBSERVATORY

BOYDEN has not always been the only major observatory near Bloemfontein. There was another: the Lamont-Hussey, on Naval Hill. Inside the great dome there was a 27-inch refractor, actually the largest telescope of its kind ever to be set up in South Africa. And yet, by tragic mismanagement, the Lamont-Hussey Observatory is now in total abeyance.

The story began in 1927, with a generous offer from R. P. Lamont of Chicago. Lamont was a rich man, and he wanted to further the cause of science. A fellow-student of his had been Professor W. J. Hussey, who was busy on a survey of northern-hemisphere double stars with the help of the 36-inch refractor at the Lick Observatory. Hussey wanted to extend his survey to the far south; Lamont agreed to provide the necessary equipment.

At this time Hussey was Professor of Astronomy in the University of Michigan, and also Director of the Detroit Observatory at Ann Arbor. Born at Mendon, Ohio, in 1862, he graduated from the Engineering School of the University in 1889. He taught mathematics at the University for two years, and took charge of the Observatory there for two years. His work on double stars dated back to his arrival at Lick in 1896; he discovered more than 1 300 new pairs, and in 1906 he was awarded the Lalande Prize of the French Academy.

As an undergraduate, Hussey had shared rooms with Lamont, who went on to become President of the American Steel Foundries. In 1910 – at Ann Arbor – work on a 24-inch refractor was begun, to be used on southern double stars. Unfortunately no suitable glass blanks for the objective could be found, and it was not possible to purchase any until after the war, when two 27-inch disks were obtained from the celebrated Zeiss firm at Jena. These were duly prepared, and in 1923 Hussey had already selected Bloemfontein as a site for the new observatory, so that everything appeared to be well in hand. His chief assistant was to be Dr R. A. Rossiter.

In the autumn of 1926 Hussey set out for South Africa to superintend the erection of the 27-inch. On 27 October he attended the meeting of the British Astronomical Association, and gave an address to its members; twenty-four hours later he collapsed and died from a heart attack – a tragic loss indeed.

Rossiter took over, and plans went ahead. Between 1927 and 1951, when he retired as Director, he discovered 7 170 new southern double stars and made a total of 28 000 measures – all very much in keeping with the traditions of the Union Observatory in Johannesburg, where van den Bos and Finsen were working along the same lines.

Attention at the Lamont-Hussey was not confined to doubles. Mars was another prime target, and at the oppositions of 1939, 1954 and 1956 Dr E. C. Slipher of the Lowell Observatory came to Bloemfontein to take photographs with the 27-inch refractor. The best compliment one can give these pictures is to say that they rivalled Finsen's. Then, from 1948 to 1951, a 10½-inch photographic telescope was brought from the Hale Observatories in California, to be used by Dr Karl G. Henize for studies of the hydrogen clouds in the Galaxy; this too was an extension of a survey which had already been carried out in the northern hemisphere. In 1962 Dr Frank Holden assumed the Directorship.

There was never any doubt about the excellence of Bloemfontein as a site, and the Lamont-Hussey Observatory had limitless potential. Yet there was one major obstacle: it belonged to Michigan, not to any South African organization. Again for reasons outside

The Lamont-Hussey Observatory on Naval Hill, Bloemfontein.
– *Courtesy of Africana Museum, Johannesburg*

the realm of science, support for it was withdrawn. Holden retired in 1973; in 1974 the Observatory was officially closed; and in 1975 the optics of the great 27-inch were taken out under the supervision of Dr Jarrett from neighbouring Boyden. Michigan handed over the building to the City Council of Bloemfontein for any use which might be considered appropriate.

So at the time when these words are being written (July 1976) the scene is one of total inactivity. The best that can be said is that the building still stands, and to renovate the 27-inch would be easy enough. To allow an institution of this type to lapse is arrant folly; let us hope that in the end a way will be found to bring the Lamont-Hussey Observatory to life once more.

11. THE NATAL OBSERVATORY: PROMISE AND FAILURE

FROM THE HUMBLE eighteenth-century beginnings, great observatories have sprung up in South Africa, and astronomers here have played leading rôles in almost every branch of research. However, there has been one outstanding failure – and an unnecessary one at that. This was the ill-fated Natal Observatory at Durban, whose one and only Director was Edmund Nevill, better remembered today by the name under which he wrote: Edmund Neison.

Neison had a remarkable career. He was born at Beverley, in Yorkshire, on 27 August 1849, and was educated at Harrow and at New College, Oxford. During the Franco-Prussian War he joined the French forces and served on Marshal Ney's staff; after the end of the war he was for a time Parliamentary reporter to the old *Standard* newspaper as well as theatrical critic. (Possibly these two rôles had much in common!) His interest in astronomy began very early. He became a Fellow of the Royal Astronomical Society in 1873, and began a serious study of the Moon, making his observations from Hampstead with a 6-inch refractor and a 9½-inch With-Browning reflector. His earliest lunar paper came out in June 1873, concerning the possibility of an atmosphere around the Moon, and in the following years he also produced other papers, some of which showed him to be an expert mathematician.

The Moon was Neison's main interest, and to set the scene it may be as well to say

(*Left*) Edmund Neison. (*Right*) Neison at the Old Natal Observatory. Neison (Nevill) had an intense dislike of being photographed, and it is probable that these are the only two pictures of him in existence. The Natal Observatory has been demolished.

something about the history of selenography. In Neison's time the Moon was just becoming "fashionable" once more, after several decades of relative neglect.

The first great selenographer was Johann Hieronymus Schröter, a German amateur who was Chief Magistrate of the little town of Lilienthal, near Bremen. Schröter's work extended from 1778 to 1813, when his observatory was sacked by the invading French, and all his unpublished observations, together with his telescopes, were looted. Next came three of his countrymen: Wilhelm Lohrmann, a Dresden land surveyor whose astronomical career was cut short by ill-health, and two Berliners, Wilhelm Beer and Johann von Mädler. Using a 3¾-inch refractor, Beer and Mädler spent years in charting the Moon. Finally, in 1838–9, they produced not only a map which was good by any standards, but also a detailed description of each important formation on the lunar surface. Logically enough, they regarded the Moon as a totally changeless world – and it is ironical that their map actually held back the progress of selenography. The general view following 1839 was that the Moon had now been fully charted, and that as it never changed there was really very little point in studying it further.

The only serious dissentient was Julius Schmidt, yet another German, who went to Greece and became Director of the Athens Observatory. Between 1840, when Beer and Mädler virtually ended their lunar work, and 1866, when events took a dramatic turn, Schmidt was really the only consistent observer of the Moon, and the map by Beer and Mädler remained the best, even though it had been compiled with so small a telescope.

Then, in 1866, Schmidt was studying the grey plain known as the Mare Serenitatis when he made, or thought he made, a startling discovery. In an isolated position on the plain Beer and Mädler, as well as other early observers – even Schmidt himself in 1843 – had drawn a deep, conspicuous crater which Mädler had named "Linné" in honour of the Swedish botanist Carl Linnæus. Yet in October 1866 Schmidt found that there was no such crater in that position; all he could see was a small craterlet surrounded by a white patch. He was confident that a real physical change had occurred, and others agreed with him; Sir John Herschel suggested that a moonquake had caused the walls of the old Linné to fall in, after which the interior had been filled up with rising lava.

This is no place to discuss the Linné controversy in detail. We know what the formation is like today, thanks to the close-range pictures taken by the Orbiter probes and by the astronauts of Apollo 15; it is a small, perfectly circular craterlet with regular walls and an interior free from débris. The balance of evidence is that there was no real change, and this was borne out by Mädler himself, who was still alive at the time; but the episode certainly re-awakened interest in lunar mapping, and telescopes all over the world were directed back toward the Moon.*

Edmund Neison took a lively interest, and he then made up his mind to revise the map by Beer and Mädler as well as producing a new description of the entire Moon. He set to work, and in 1876 he published the first great lunar book to be written in English. It was entitled *The Moon, and the Conditions and Configuration of its Surface;* it became a classic, and established Neison as the leader in his field. He was a founder member of the newly-formed Selenographical Society, and became its Secretary (indeed, the decline of the

*For a full discussion of this fascinating problem, see *Guide to the Moon* by Patrick Moore: Lutterworth (Britain) and Norton (New York), 1976.

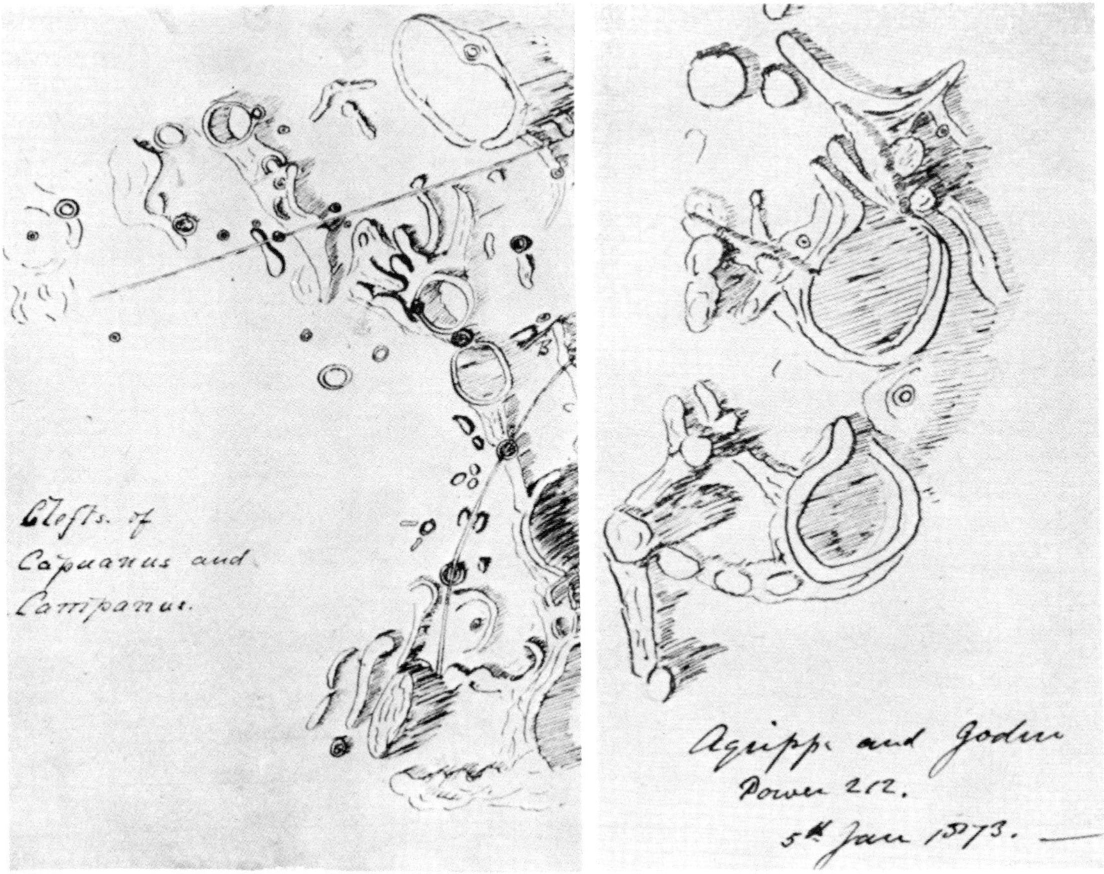

Lunar sketches by Edmund Neison of Natal.

Society began with his resignation in 1883). With C. T. Kingzett, he organized a meeting which resulted in the formation of the Institute of Chemistry, and he also wrote technical papers on optics. Then came an event which transformed his whole career: a transit of Venus.

Since Venus is closer to the Sun than we are, it can occasionally pass between the Sun and the Earth, so that it then appears as a black spot in transit against the brilliant solar disk. Because of the orbital tilt of the planet, this does not happen often. Two transits occur, separated by eight years, after which there are no more for over a century; thus there were transits in 1761 and 1769, 1874 and 1882, and the next will take place on 2004 and 2012. (Incidentally, Captain Cook's expedition to the South Seas, during which he discovered Australia, was originally dispatched so that he could take a party of astronomers out to a suitable site for observing the 1769 transit.)

The significance of transits of Venus was that they could provide a means of measuring the length of the astronomical unit. Before the present era of space-probes and radar, the transits were of immense importance, and every effort was made to observe them. As the whole method is now obsolete there is no point in describing it further, and the transits of 2004 and 2012 will not be regarded as of more than academic interest, but the situation in Neison's time was very different. Results from the 1874 transit were not wholly satisfactory, for various reasons, but more was expected from the transit of 1882, which would, it was thought, be well seen from South Africa.

David Gill took up the problem. It had long been planned to set up a new observatory at Durban. It was decided to make it ready for the transit, and in June 1882 the Durban Corporation voted a sum of £350 for the purpose; another £500 was added by the Legislative Council in the following July. An 8-inch refractor was purchased by H. Escombe, who had been working together with Gill on the project, and the Government provided a 3-inch transit instrument.

Gill wanted a skilful and permanent Director; and who better than Edmund Neison? There was need for haste. By the time that he was ready to issue an invitation it was already June 1882, and the transit was due in December.

Gill sent Neison an urgent telegram, offering him the post of Government Astronomer at Natal, and urging him to arrive in time to observe the transit. It was a challenge which Neison did not feel disposed to ignore. He accepted, packed up, and sailed for Durban, arriving on 27 November 1882. By 1 December he was already at the Observatory, only to find that things were far from ideal; for instance, the machinery for moving the dome had been carefully covered with a thick layer of paint, while the polarizing eyepiece supplied by the firm of Merz, while doubtless excellent in quality, suffered from the slight disadvantage that it would fit neither the telescope nor its accessories! The transit instrument was in Cape Town, and could not be shifted in time. Neison had to do some frantic improvisation, and in the end he observed the transit with great success.

It was an encouraging start. The Natal Observatory had begun its work, and Neison made great plans. He intended to improve tables of the Moon's motion, carry out general observations in collaboration with the Cape, and also to set up a meteorological station. Publications were put in hand, and all seemed to be in order. Alas, the story from then on was one of perpetual frustration.

The trouble was, quite simply, lack of funds. Neison had been promised an adequate salary, with provision for assistants, and also grants to help with the Observatory's publications. As so often happens, the authorities broke their word. Publications were prepared – and remained on the files; there was no money to print them. After 1887 there was no assistant, and Neison had to depend on sporadic help from local amateurs. He did have a new assistant for a while from 1888; a manuscript catalogue of the right ascensions of Zodiacal stars was prepared, as well as tide-tables for Natal, and Neison continued to do his best. He also became Government Chemist and Official Assayer for Natal, acting sometimes as pathologist in cases of suspected poisoning!

It was a noble effort, but as time went by things became more and more difficult. Money was not forthcoming, and publications from the Observatory had to stop, so that the records remained in manuscript form. Eventually even Neison had to admit that the odds against him were too great. Funds were cut off altogether, and in 1911 the Observatory was closed down. Neison left South Africa and returned to England; he settled in Eastbourne, and lived there in retirement. His astronomical work was more or less over, though he kept up his scientific interests, and was awarded the Medal of the Royal Chemical Society in 1935. He died at Eastbourne in 1940.

Quite apart from astronomy, Neison had many interests. One was sport. He was an excellent tennis-player, and it was he who introduced lawn tennis to South Africa; his wife, née Mabel Grant, whom he married in 1894, was South Africa's tennis champion for eleven years. He was also a good golfer. He was extremely skilful at sketching, in a general way as well as astronomically; he was intensely interested in Babylonian history, upon which he collected a vast store of information which remains unfortunately unpublished. He also wrote some novels, but these too remain in manuscript form, since they were never submitted to a publisher. In many ways he was shy; he twice declined an invitation to become a Fellow of the Royal Society, though finally, in 1908, he accepted. And he hated being photographed; the admittedly rather indifferent print reproduced here is the only one in existence which shows him at all clearly.

His marriage was extremely happy. His wife survived him by some years; of his three children one of his sons died of war wounds, and the other now lives in South Africa, while his daughter, Miss Maud Nevill, lives in London.

The story of the Natal Observatory is not a happy one. Had Neison been given even a fraction of the support he had been promised, he would have accomplished far more than he was actually able to do. Nothing now remains of the Observatory, and it must be admitted that a great opportunity was lost. Yet Neison himself will always be remembered, and nowadays, when we are used to pictures of astronauts driving around the surface of the Moon, Neison's book of 1876 takes us back into another and more placid age. This is inevitable; present-day maps are compiled by using every modern technique – plus colossal sums of money, whereas Neison's work was undertaken on his own. It cost nothing in hard cash, but it provided a firm basis for all later lunar studies, so that in a way it provides a link between the past and the present. It is tragic that in South Africa he was unable to fulfil the tasks for which he was so clearly fitted. The fault was not his.

12. INTO DEEP SPACE: THE GREAT DISH AT HARTEBEESHOEK

On 4 October 1957, the Space Age began. It started not with a whimper, but with quite a considerable bang. Without any definite prior announcement, and to the surprise (plus, it must be admitted, consternation) of many Western scientists, the Russians launched the first of all artificial satellites, Sputnik 1. It was a midget probe, not much larger than a football, and carrying little apart from a radio transmitter; but it marked the start of a new era, and few people who heard its famous "Bleep! bleep!" signals will ever forget them.

Reactions were mixed. True scientists expressed gratification; not many agreed with Rear-Admiral Rawson, of the United States Navy, who stated that the Sputnik was "a hunk of old iron that almost anyone could launch". At that time, indeed, the Americans were unable to launch an orbital piece of iron or anything else. Their first success came in the following year, with the launching of Explorer 1 by the team led by Wernher von Braun. Then, as month followed month, techniques were improved, and man too entered space; the first flight was made by Yuri Gagarin in 1961, and less than nine years later Neil Armstrong and Edwin Aldrin stepped out on to the surface of the Moon.

South Africa has as yet taken no part in the programme of manned space-flight, but unmanned probes come very much into the story, because as recently as 1974 the great radio tracking station at Hartebeeshoek, 40 miles north-west of Johannesburg, was a vital link in the chain.

With an Earth satellite, the vehicle itself is taken up by a launching vehicle and put into a closed path round the world. Provided that it is high enough to avoid dipping into the denser part of the atmosphere, it will remain aloft indefinitely, and will move in precisely the same way that a natural body would do. This, after all, is how the Moon goes round the Earth, or the Earth round the Sun. Only if the probe spends any appreciable part of its elliptical orbit at a lower level, where it has to contend with friction against the atmosphere, will it spiral downward and decay. This was the fate of Sputnik 1, as well as many other early probes; we have already noted how the observations and predictions made by Finsen at Johannesburg were of extreme importance. If the probe is at a safe altitude, it will be permanent. By now, of course, thousands of Earth satellites have been launched, fulfilling rôles ranging from weather studies to telephone, radio and television relays.

If the probe is destined to leave the region of the Earth altogether, and go out to another world, it has to be tracked; and in general, mid-course corrections are needed. The Moon is comparatively close, but the planets are not; Venus, the nearest of them, is always at least a hundred times as far away as the Moon, while Mercury and Mars are even more remote. Beyond Mars, and beyond the asteriod zone, come the giant planets, beginning with Jupiter, which cannot be reached in less than about eighteen months even with our latest probe vehicles. All these planets were by-passed in the 1970's, and by 1976 Pioneer 11 was even on its way to distant Saturn. But in the early days things were much less efficient, and one essential was to keep each probe in touch with tracking stations on the surface of the Earth.

The central organization was, of course, NASA – the National Aeronautics and Space

Administration, with its headquarters at Houston in Texas, and its main launching ground at Cape Canaveral in Florida. (The name was changed to Cape Kennedy during the 1960s, but the alteration was not popular, and eventually it was changed back again.) One facility of NASA was the Deep Space Network or DSN, which was in turn divided up into several departments; among these was the Deep Space Instrumentation Facility or DSIF, whose rôle was to deal with space communications and the various tracking stations.

It was this which brought South Africa into the organization. Obviously, no single tracking station can follow a probe continuously; if the vehicle is above the horizon at, for instance, Houston it will be below the horizon from the other side of the world. Continuity was vital at first, because it was impossible to forecast the movements of satellites accurately for more than a few hours ahead; there were too many uncertain factors, such as the imperfectly-known density of the Earth's upper air. Therefore, NASA established a whole series of tracking stations, separated by an average of 120 degrees in longitude. Among them were Woomera and Tidbinbilla in Australia, Ascension Island in the South Atlantic, Robledo and Cebreros in Spain, Goldstone in California, and, of course, Cape Canaveral in Florida. The chain was completed by the station at Hartebeeshoek. There had to be a South African station; otherwise the chain would have had an intolerable gap.

Accordingly, the Government of the United States and South Africa entered into an agreement according to which Hartebeeshoek would become an official part of the DSIF. Work was started in January 1961, and in the following July – an amazingly short time – the Johannesburg Deep Space Station was in full order; there were no major delays, and it was arranged that the South African CSIR should be responsible for the actual operating. As a matter of fact, the Station was only marginally concerned with Earth satellites after its initial period, and its main function was in tracking the probes which went further afield. There were the lunar probes; the Rangers which crash-landed on the lunar surface, the Surveyors which made soft landings, and above all the Orbiters, which were put into closed paths round the Moon and sent back photographs covering the whole surface from close range. (Without the Orbiters, the manned Apollo programme could never have been risked.) Then there were the Mariners to Mars, to Venus and subsequently to Mercury; two Pioneers to Jupiter, and various assorted probes which had varied purposes. The results were sent to NASA, and everything was pooled to the benefit of all.

By the time that the tracking of deep-space probes had been fully developed, Hartebeeshoek's original "tracking dish" had been augmented by a huge radio paraboloid with a diameter of 85 feet, as tall as a ten-storey building and with a total weight of about 600 000 pounds. It was – and is – a most impressive structure, visible from miles around and dominating the landscape. It remains much the largest fully-steerable radio telescope in South Africa, and one of the largest in the southern hemisphere, and it can be controlled with remarkable precision, so that it can cover the whole of the sky.

There is no essential difference between a radio wave and a lightwave, except in length. Visible light is of extremely short wavelength, generally measured in units known as Ångströms in honour of the last-century Swedish scientist Anders Ångström. One

The 85-ft (26 metre) Great Dish at Hartebeeshoek.

– Dr Patrick Moore

Ångström is equal to a hundred-millionth part of a centimetre. Red light has a wavelength of about 7 500 Ångströms, which is not very much; blue light goes down to only 3 900 Ångströms. If the wavelength of the radiation lies outside these limits, our eyes are not affected. To the short-wave end we have the ultra-violet, X-rays and the incredibly short, penetrating gamma-rays; to the long-wave end we come to radio emission, which can be collected by the instruments known as radio telescopes even though they are really more in the nature of huge aerials. The Hartebeeshoek dish is one of these, and so was the now disused smaller dish nearby, for many years operated on behalf of NASA's Space Tracking and Data Acquisition Network.

The dish itself, with an area of one-eighth of an acre, is attached to a special quadropod support, which itself holds the gold-coated secondary or sub-reflector, below which is a cone. The radio waves from space hit the dish and are sent up to the sub-reflector; they then pass through wave-guides into the cone, which contains the receiver. The signals then pass through complicated systems of circuits and end up in the control room in the main building, several hundred yards away, where the results are obtained in the form of a pen-trace on a paper attached to a revolving drum. Radio astronomy has become of fundamental importance in astronomy as well as in probe tracking; events have moved quickly since the American radio engineer Karl Jansky accidentally happened upon radio waves from the sky less than half a century ago.

All through the first decade and a half of the Space Age, Hartebeeshoek operated without a break. Its most distant target was Pioneer 10, the first Jupiter probe, which by-passed the Giant Planet in December 1973, narrowly escaping being put out of action by the intense Jovian radiation. But then came a change of emphasis. Yielding to political pressure, the NASA authorities decided to give up the Johannesburg Deep Space Station, and tracking there ended in June 1974.

The Hartebeeshoek staff, under Dr George Nicolson, had to revise their programme. Previously they had been concerned mainly with the probes but they had also carried out a certain amount of research in conventional radio astronomy. With the NASA defection, they had either to close down or else extend the radio astronomy programme. Fortunately, funds were sufficient; and from being a tracking station, Hartebeeshoek emerged as a fully-fledged radio astronomy observatory. The decision was certainly wise in view of the excellence and power of the great 85-foot dish.

The change-over did not take long. The NASA equipment was taken away, and the secondary dish, built exclusively for tracking, was put into mothballs. The stage was set for an all-out onslaught upon purely astronomical problems.

When Jansky first picked up cosmic radio waves in 1931, with his home-made aerial, he soon found that they came from the Milky Way; but it was not until after the war that radio astronomy came into its own, and very large instruments were constructed. (Not all radio telescopes are alike. Some consists of long lines of aerials, while others look more like collections of barbers' poles!) It was found that although some relatively close objects are radio emitters, including the Sun and Jupiter, most of the sources are very remote. In our Galaxy, they take the form of the remnants of supernovæ or exploded stars, of which the most famous example is the Crab Nebula in Taurus; we know that this expanding gas-cloud is the débris of a star which was seen to blaze out in the year 1054, becoming so brilliant that for some months it was visible in broad daylight. But our

Galaxy is only one of many, and most of the interesting radio sources lie in external systems.

Galaxies themselves are of many kinds. Some, such as the Clouds of Magellan, are irregular; others are spiral, others elliptical. Not all of them are strong radio emitters, but there are some galaxies which are extraordinarily powerful in the radio range. The cause of this radio energy is still uncertain, and the first essential is data-collection, which is one of the main programmes now going on at Hartebeeshoek. However, even greater emphasis is being placed there on the enigmatical objects known as quasars – a convenient abbreviation for "quasi-stellar objects".

When catalogues of radio sources were drawn up, it was found that most of the sources coincided with supernova remnants or with galaxies. In 1963, however, it was found that one source was in the same position as an object which looked like a dim bluish star. Optical astronomers were called in, and provided a major sensation by announcing that the object was not a star at all; it was something infinitely more dramatic – the first quasar. It lay at a distance of thousands of millions of light-years, and was presumably much more powerful than an average galaxy even though it was far smaller. Other quasar discoveries followed, and it now seems that a powerful quasar may outshine a normal galaxy by a factor of at least 100. Moreover, the quasars are receding from us at tremendous speeds – in some cases over 90 % of the velocity of light.

Quasars are as puzzling today as they were when first identified. They may be the nuclei of exceptional galaxies; they may be entirely different in nature, shining by some process as yet unknown; it is even possible (though unlikely) that there has been a major miscalculation, so that they are not so remote or so luminous as most astronomers think. As always, observational data are needed, and this is where Hartebeeshoek is playing an exceptionally important rôle. In particular, emphasis is being placed on the short-term changes in the brightness variations of quasars, which are quite marked in both the optical and in the radio range.

To gain full information we would need to use a radio telescope of immense size – perhaps as large as the Earth itself. Since this would pose some rather difficult problems, an alternative has to be found, and the solution is to use two or more radio telescopes which are widely separated (generally in different continents). The signals from the quasar under study are recorded simultaeously by both instruments. The results are then processed by combining the tapes, and very fine resolution can be obtained; it is equivalent to looking at a man walking about on the surface of the Moon, or an observer standing in Johannesburg and watching a bee going about its lawful business in Cape Town.

Obviously there are many problems to be solved. Hartebeeshoek, with its huge radio telescope and its ultra-modern auxiliary equipment, has much to contribute in the coming years, and in this sphere there is full, friendly collaboration between South Africa and the other scientific nations of the civilized world.

13. TEMPORARY OBSERVATORIES

We have noted that there are some South African observatories, of great repute, which are now inactive or definitely defunct. The Radcliffe Observatory is the main case of a great institution which has been destroyed as effectively as the Romans disposed of Carthage (though admittedly for rather different reasons). As professional establishments the Republic Observatory in Johannesburg and the Lamont-Hussey in Bloemfontein are at best in abeyance, though they still stand, and it would be premature to say that there is no chance of their rising, phœnix-like, in the future. The Yale Observatory operated for many years, but now it too has gone. In addition there were various observatories which were never intended to be more than temporary, and it would be wrong to omit them from any history which set out to be reasonably comprehensive. Actually, Father Tachard's observatory comes into this category, but was definitely a special case, so let us pass on to modern times.

First, there was the solar observatory at Mount Brukkaros, sixty miles north-west of Keetmanshoop in South-West Africa, which was set up by the Smithsonian Institution in 1926 and was financed by the Royal Geographical Society in London. It had one purpose, and one only: to measure the solar constant – that is to say, the total amount of solar radiation received on the Earth. This may sound a simple task, but in fact it is nothing of the kind, as generations of astronomers have found out. Mount Brukkaros was thought to be a suitable site, and with the full co-operation of the Government of South-West Africa the project was put under way. Altogether, the station was functioning from October 1926 to June 1932.

Many measurements were made, but it cannot be said that the results came up to expectations. The trouble was, simply, dust – stirred up by the high winds and left suspended in the upper atmosphere. These were precisely the conditions which the observers did not want, and eventually they gave up.

Two years later, in 1934, came a purely German team, organized by the Observatories of Breslau, Neubabelsberg, Sonneberg and Bamberg. They settled at Windhoek, bringing with them short-focus, wide-aperture cameras and a fast spectrograph. They had two aims. One was to study the large-scale structure of the Milky Way, make estimates of variable stars, and investigate those two elusive sky-glows, the Zodiacal Light and the Gegenschein. The second was, apparently, to see whether Windhoek would be suitable as a site for a permanent German station. They decided against it; and though a Sonneberg team came again for a brief period in 1952-3, there has since then been no professional astronomical observatory in South-West Africa. (Neither are there any official observatories in Rhodesia or in neighbouring countries. In 1965 Dr J. M. Bacher, Professor of Physics at the University of Botswana, put forward ideas for an observatory in Lesotho, and building was actually started; but funds were lacking, and nothing further was done.)

In 1953 various authorities in Belgium, France, Holland, Sweden and West Germany discussed possible sites for a completely new Southern Hemisphere Observatory to be run on international lines, and South Africa, with its fine climate, was an obvious possibility. In 1955 two "pilot" observatories were set up; one by the French at Zeekoegat, near Prince, and one by the Germans in the Newveld Mountains, west of Beaufort West.

Steve Heninger and Patrick Moore beside the Laser-Transmitter at Olifantsfontein.
– *Pete Collins*

The German station soon closed down. Zeekoegat continued for longer, but was no more than a testing site, and for reasons mainly unconnected with science South Africa was abandoned as a site; subsequently the Observatory was set up in Chile.

Last, but by no means least, there was the special installation at Olifantsfontein, in the Transvaal. This was a station set up by NASA entirely for the observation of Earth satellites. It began work in 1959, with a Baker-Nunn camera – an instrument developed specially for this purpose; it recorded satellites photographically, and measured their positions with surprising accuracy. Later came a laser transmitter, in which a beam of concentrated light was reflected from the target satellite; the results enabled the observer to measure not only the satellite's position, but also its distance and movement. For most of the time the main observer was Steve Heninger; with the laser, the satellites studied were Geos I and Geos 2, both of which were American-launched and had been sent up for geophysical purposes. Eventually, and again mainly because of political pressure, the Olifantsfontein station was given up. It was dismantled in 1976.

So much, then, for the short-lived observatories. Apart from Olifantsfontein, it cannot honestly be said that they accomplished much, but from an historical point of view it is only right to put them on record.

14. ASTRONOMY IN THE UNIVERSITIES

It is axiomatic that all major universities should concern themselves with astronomy in some way or other. Naturally, some are more involved than others. In South Africa, for instance, the Department of Astronomy at the University of the Orange Free State has been more or less synonymous with the Boyden Observatory, and there is no need to say more about it separately. When we come to the Cape, however, things are different. Of course the University authorities make full use of the facilities at Sutherland, just as they used to do at the old Cape Observatory itself, but there is a distinct character to the Department of Astronomy, which is headed by Professor Brian Warner.

Warner hails from Southern England, from the neighbourhood of East Grinstead in Sussex.* After graduating from London University he went to the Radcliffe Observatory to work for his doctorate; then, after a spell at the University of Austin in Texas, he came back to Cape Town as Professor of Astronomy in the University. He has carried out a good deal of pioneer work in astrophysics, but is perhaps best known for his researches into the strange stars known as dwarf novæ.

True novæ are stars which show sudden, cataclysmic outbursts, flaring up to become prominent for a few days, weeks or months before fading back into obscurity. Some of them become really brilliant, and plenty of novæ have been seen during our own century – a recent example being Nova Cygni of 1975, which for a brief period reached magnitude $1\frac{3}{4}$, though it was too far north to be well seen from South Africa.

It is important to "catch a nova in the act", so to speak, because studies of the rapid development during the outburst are of tremendous theoretical importance. Amateurs have a fine record of nova discovery. The trouble is that they cannot be predicted; there are myriads of faint stars in the sky, and to keep watch on them all is absolutely out of the question. What can be done, however, is to study novæ during their decline, and after the outburst is definitely over. It has been found that most, if not all, are binary systems, of which one component is a very small, super-dense star of the type known as a White Dwarf. Such is the old nova which flared up in 1925, and is now called RR Pictoris. The first evidence of its double nature was found by Finsen with the $26\frac{1}{2}$-inch reflector at Johannesburg.

This is no place to go into details about stellar evolution; suffice to say that a White Dwarf is a very old star, which has exhausted its nuclear energy and has collapsed. Many are known, the most famous example being the dim companion of Sirius. It seems a general rule that one component of an old nova binary is a White Dwarf, and this is the case with another bright nova of our own century: DQ Herculis, which flared up to the first magnitude in 1934, but has now become very faint. Twenty years after the outburst, M. F. Walker in America found that the nova was fluctuating very quickly in a period of 71 seconds, and there was an even quicker "flickering". This behaviour had not been

*I lived in East Grinstead for thirty-six years, and during Warner's school days he and I spent many hours together observing the Moon together from my private observatory. Subsequently he became Director of the Lunar Section of the British Astronomical Association; when he left for South Africa in 1963, I succeeded him. I continue to be a lunar observer, but Warner has now moved on to greater things! – P.M.

Professor Brian Warner explains a dwarf nova outburst.

– Dr Patrick Moore

noticed earlier because the amplitude is very low; to measure very small changes in brightness means using photoelectric equipment, together with a telescope large enough to collect sufficient light.

Associated with novæ are the variable stars which suffer smaller but much more frequent outbursts. They are sometimes called SS Cygni or U Geminorum stars, and sometimes dwarf novæ. All seem to be close binaries, one component being a White Dwarf while the other is a larger, cooler star. Because the White Dwarf is the hotter and brighter member of the pair, it is usually termed the primary, though it is much smaller than the cooler, redder secondary.

For a long time it was uncertain whether the frequent outbursts took place in the White Dwarf primary, in the secondary, or in a shell of gas surrounding the whole system. It now seems that the White Dwarf is the culprit, but there are pulsations and flickerings too. (These have nothing to do with the visual twinkling of a star as seen with the naked eye or in a telescope. Twinkling of this sort is due purely to the Earth's atmosphere, and is unconnected with the star itself.) The flickerings of dwarf novæ are intrinsic, and are too slight to be noticed visually.

At the Cape University, Warner has developed equipment which enables him to study these very slight variations. The actual observations are made from Sutherland, and computers at the Department of Astronomy at the University do the rest – which sounds easy, but is nothing of the kind. From this work it has been possible to build up a picture of a dwarf nova system. Normally the components are so close together that they practically touch, and are whirling round their common centre of gravity very rapidly indeed. The mutual tidal forces between the stars are extremely strong. The less massive secondary is the greater sufferer, and material is pulled away from it, moving across to the primary in a narrow stream and producing a disk or shell of gas round the White Dwarf. Where the stream of material hits the disk, a bright spot is formed – so bright, indeed, that it out shines both the components put together. The flickering originates in the bright spot, and is due to irregularities in the stream of gas coming from the secondary; the pulsations, of rather longer period and greater amplitude, are caused by outbursts in the White Dwarf itself.

This particular piece of research has been described because it is typical of the work now being carried out at University level. Other astronomers at the Cape University are engaged in research of various kinds, and in addition there is laudable concentration upon the popularization of astronomy; the various public lectures are always well attended. There is also a small planetarium, not at the University itself but in the South African Museum.

Planetaria are educational establishments, and have become more and more popular during the past few years. The modern-type planetarium dates back to the work of Professor Walther Bauersfeld, of the Zeiss Optical Works in Germany, just before the first world war. Basically, a complicated projector throws an "artificial sky" on to the interior of a large dome; the star patterns are faithfully reproduced, and there are special projectors to deal with moving bodies such as the Sun, Moon and planets. In a major planetarium it is sometimes difficult for a member of the audience to realize that he (or she) is not sitting in the open, looking at a real sky.

The advantages of a planetarium are many. Identification arrows can be used to help in star recognition; features such as the celestial equator and poles can be projected; and, of course, the movements of the planets and other members of the Solar System can be speeded up to demonstrate how they really behave. Moreover, a planetarium can show the sky as it would appear at any date and time from any part of the world, and a skilful lecturer can impart an amazing amount of information in a display lasting for no more than an hour or so.

The Cape Planetarium has a seating capacity of seventy, and a projector built by the Spitz Company in America. The Planetarium at the University of the Witwatersrand, in Johannesburg, is larger and more elaborate, with a Zeiss projector and a 67-ft. dome. It has been in regular use for many years; when it was founded, in 1959, the Director was Professor Arthur Bleksley, who is well known all over South Africa for his lectures and for his radio broadcasts. Bleksley was succeeded by the present Director, Thomas Geary, under whom the Planetarium continues to flourish. Though largely autonomous, it is an integral part of the University, and as such must be included in the present chapter.

Another topic which cannot be omitted, even though it may seem to be something of a digression, is that of the Vredefort Ring, which has been studied by members of the Bernard Price Institute of Geophysical Research at the University. The Ring itself is a circular

The Planetarium at the University of the Witwatersrand, Johannesburg.
– *By courtesy of the Johannesburg Planetarium*

structure, 30 miles in diameter, some way from Pretoria and just inside the Orange Free State; the town of Parys lies inside it, as well as the village of Vredefort from which the structure takes its name. It must be mentioned here because there have been suggestions that it is of impact origin: that is to say, produced by the fall of a large meteorite in prehistoric times. The same origin has been claimed for the principal craters of the Moon, which takes us back to pure astronomy.

Meteorites are not uncommon, and most museums and universities have collections of them. Now and then we find a truly giant specimen, and at present the known holder of the heavyweight record is the Hoba West Meteorite, which fell near Grootfontein in South-West Africa – again presumably in prehistoric times. It is still lying where it fell, and it is questionable whether anyone will try to run away with it, since the estimated weight is about 60 tons. It made no crater; but there is no doubt that the famous Arizona Crater, not far from the Lowell Observatory at Flagstaff, is meteoritic.

Other impact craters are known, though in some cases the origin of the structures is debatable. So far as the Moon is concerned, there are two main theories of the genesis of the craters: they may be meteoritic, they may be volcanic – probably both processes have operated.* In either case, there may be some analogy with the Vredefort Ring.

*One of the present writers (P.M.) is particularly interested in the Moon, and is strongly of the opinion that vulcanism has been the main agent in the moulding of the lunar surface; but this is a personal view, and others will certainly disagree.

The Hoba West meteorite.

– Ludolf Meyer

The Ring has been known for many years, but the most intensive research is modern, and was started in 1961 by Professor L. O. Nicolaysen and his colleagues of Wits University. It is an interesting structure; the "walls" are very evident to anyone who drives up to the Ring from the direction of Pretoria, though in places they are low and incomplete. The latest researches, many of which have been undertaken by Duncan Stepto of the Bernard Price institute, strongly suggest that it is of volcanic rather than impact origin. Certainly it is of distinctly "lunar" aspect, and it provides yet another example of the intimate connection between astronomy and the earth sciences.

Rhodes University, at Grahamstown, also has a flourishing Department of Astronomy, but here the main emphasis is upon radio studies. The long-wavelength emissions from Jupiter have been carefully examined with a special radio telescope which is quite undishlike, and is of the "barber's-pole" variety; lately, too, the Rhodes workers have been making use of the Hartebeeshoek equipment for their studies of the dust and gas clouds between the stars of our Galaxy.

These are but a few of the many researches undertaken by University departments; and, needless to say, it is the universities which provide the astronomers of the coming years. Without them, the future of astronomy would be bleak, but as matters stand at the moment there need be no misgivings.

15. THE AMATEURS

As a science, astronomy is unique in one respect: amateurs can not only take part, but can play a really major rôle. The enthusiast who sets up a modest telescope in his back garden can make himself extremely useful, and may even become world-famous.

Before the Space Age, mapping the Moon was carried out chiefly by amateurs. Earth-based lunar research still goes on; for instance, it is the amateurs who are taking the lead in studying the occasional brief glows which indicate a certain amount of residual activity. Jupiter, with its ever changing surface, is always a favourite object, and the famous white spot on Saturn in 1933 was first seen by a most unusual amateur, W. T. Hay, who will be better remembered nowadays as Will Hay, the stage and screen comedian (incidentally, a personal friend of both the present writers). Most significantly of all, perhaps, amateurs search for new comets and novæ. In England, G. E. D. Alcock now has four comets and four novæ to his credit, while another English amateur – J. P. M. Prentice – was the discoverer of the important 1934 nova, DQ Herculis.

South Africa has produced many gifted amateurs. One, of special eminence, was Alexander William Roberts, who was born in Scotland in 1857, but at the age of twenty-five came to South Africa with the aim of devoting his life to education. He went to Lovedale, a Black college near the village of Alice, where he stayed for the rest of his life apart from one return visit to Scotland in 1897 to recuperate and rest. At first his quarters in Lovedale were decidedly cramped, and it is said that his wife had to cook their meals in a three-legged pot over an open fire.

Roberts' official work was in teaching, and he carried out his duties with the utmost conscientiousness, but he was also a fanatical observer, who had been introduced to astronomy by no less a person than Sir David Gill. He was interested mainly in variable stars, and as soon as possible he erected a small observatory, equipping it with a pair of field-glasses and a second-hand theodolite! Working systematically, he would examine a set region of the sky, and return to it a few nights or weeks later to see what changes had taken place. In this way he discovered twenty new variables, four of them of the Algol type. With an Algol star, the fluctuations are due not to intrinsic changes, but to the eclipse of the main star by a less luminous companion; the prototype is Algol or Beta Persei, which is a prominent naked-eye star in the northern hemisphere of the sky.

Roberts became fascinated by Algol variables, and published some important papers about them. He then acquired two special telescopes, with which he could bring a variable and a standard comparison star into juxtaposition. With this equipment it is said that he could estimate the brightness of a star with an error of no more than 2/100 of a magnitude, which is precision indeed.

Some of his work was outstanding. For instance, he discovered the variability of the star now known as RR Centauri, and made no less than 10 000 estimates of it. Eventually he was able to give the following description of it: "This variable is in some respects the most remarkable of all the apoidal binary pairs. They coalesce, forming a dumbbell. The prolateness of both components is much greater than that of any other star of the same class. The light-curve indicates a slightly elliptical orbit. The density is 0.25, and there is evidence of tidal contraction and expansion."

Dr Patrick Moore and Jack Bennett, the Pretoria Comet Hunter.

– Pete Collins

Roberts did not confine himself to Algol stars. He paid equal attention to the Cepheids, and in particular he searched for any steady, permanent decreases in their amplitudes. Then there were the long-period variables, for which he wrote that "in the case of over seventy of these stars there is the clearest evidence of a period-variation usually from 20 to 30 times the length of the mean period". Unfortunately he did not have sufficient time for the full reduction and analysis of his observations. Throughout the day he was teaching; at night he observed, usually having no more than three hours' sleep. He overworked chronically, and his health suffered. Moreover, he fought a lone battle from an isolated outpost, and at the time of his death, in 1938, his observations were still not properly sorted out. This was tragic; but in spite of the difficulties he faced, Roberts accomplished more than any one man might reasonably have been expected to do.

Roberts, with his quarter of a million accurate observations of variable stars, set a standard which was hard to equal, but he has had worthy successors. Several South African amateurs have been responsible for comet discoveries; W. Reid, A. F. I. Forbes and R. P. de Kock of Cape Town, for example, and H. E. Houghton and G. E. Ensor of Pretoria. (Ensor's Comet of 1925 was something of a disappointment. It showed signs of becoming quite bright, but as it approached the Sun it became diffuse and faded out; it was never seen again, and presumably disintegrated in the way that these wraith-like objects sometimes do.) And in our own time there is Jack Bennett, also of Pretoria.

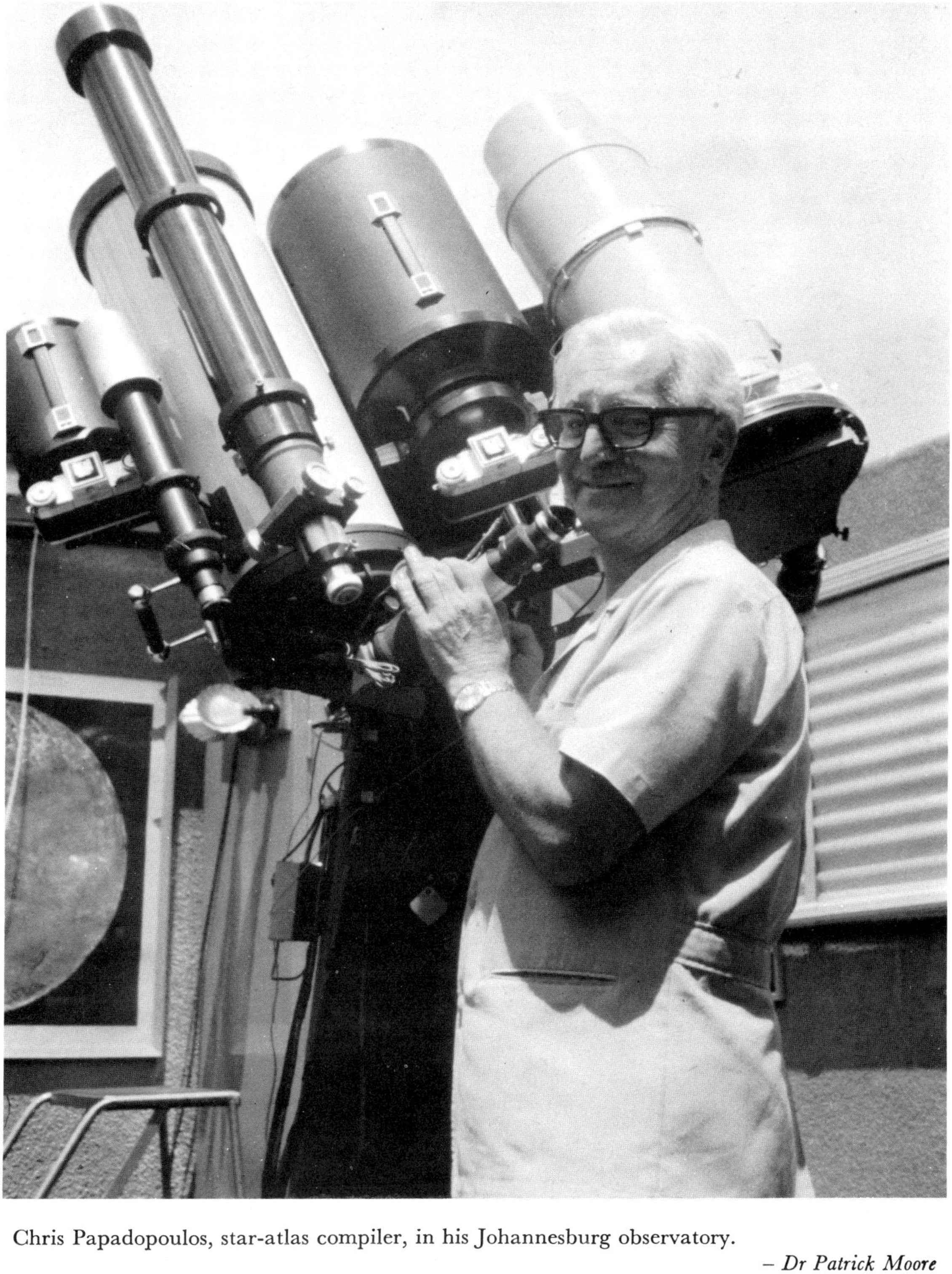

Chris Papadopoulos, star-atlas compiler, in his Johannesburg observatory.

– *Dr Patrick Moore*

Jack Bennett, too, is a comet-hunter. Originally, in the years immediately following the launching of the first Earth satellite in 1957, he was a member of the Moonwatch patrol – a band of observers, both professional and amateur, spread all over the world, and whose task was to keep visual track of the probes which had been put into paths round the Earth. For this work Bennett used a modest portable telescope with a wide field, kept in his garden – or, to be more precise, in an outside toilet; moving it into the courtyard and making it ready for use took only a minute or two! Gradually he turned over to comet-hunting, and he used the traditional method of sweeping across the sky time and time again, checking the stars and doing his utmost to identify anything unusual.

The work is far from easy. The first essential is to learn the star patterns perfectly, which takes years; remember, even a small telescope increases the number of visible stars tremendously. Then there is the time factor. New comets are not common, and to locate one means many hours of searching. Bennett usually sweeps for at least 100 hours in each year, and so far his name has been attached to two comets, one of which – the comet of 1970 – became brilliant; it was in fact the brightest visitor for many years, its only more recent rival being West's Comet of 1976. Bennett is also the only living astronomer to have made a visual discovery of a supernova in a remote galaxy. In his own words: "I was looking for comets, and I found an object which seemed peculiar because it had a star in the middle which ought not to have been there . . . I made a few tests, looked up a few books, and telephoned the Radcliffe Observatory." Within a few hours the professionals confirmed that it was indeed a supernova; and thanks to Bennett's advance warning, they were able to take some vital spectrograms of it while it was at its peak.

Of course, not all amateurs are concerned with comets. One who is not is Christos Papadopoulos, of Johannesburg, who came to South Africa from his native Greece many years ago; by profession he was a civil engineer, and after his retirement he took up astronomy in earnest. From 1960 he has observed regularly, mainly by means of photography; he has built an observatory containing a magnificent 16-inch photographic reflector, and his results are up to any professional standard. His planetary photographs are outstanding, but his main work to date has been the compilation of a photographic star atlas covering as much of the sky as is conveniently accessible from Johannesburg. The work was started in 1968, and was complete by 1976; when published, the Atlas will be of lasting value to astronomers all over the world. Papadopoulos has also played a great part in the development of amateur astronomy in Johannesburg, and it was fitting that in 1975 he should become President of the Astronomical Society of Southern Africa.

The ASSA was founded in 1922, as a result of the amalgamation of the Cape Town and Johannesburg societies. In point of time the Cape came first; it originated in 1910, largely because of the interest aroused by the return of Halley's Comet. Johannesburg followed in 1918, and in 1932 the Natal society joined the main body. A regular journal is published, and today there are local centres in the Cape, Durban, Pietermaritzburg, Bloemfontein and Pretoria; the Port Elizabeth branch is dormant (only temporarily, it is to be hoped). The main emphasis is on the observation of variable stars, comets, meteors, and occultations of stars by the Moon – particularly "grazing occultations", during which the star brushes past the Moon's limb. A leading specialist in this field is Danie Overbeek of Johannesburg, who has his own well-equipped observatory, and who in 1976 followed Papadopoulos as President of the ASSA.

David Block in his observatory at Krugersdorp.

– Dr Patrick Moore

All the local Centres have their own observatories, equipped with telescopes which are powerful enough to be really useful; they hold regular meetings, and do a great deal toward encouraging a general interest in astronomy. There is also the point that some current professional astronomers of true eminence began their careers as amateurs. Professor Brian Warner is one. Another is David Block, who erected an observatory at his home in Krugersdorp when still a schoolboy; he was elected a Fellow of the Royal Astronomical Society at the age of nineteen, and before he had celebrated his 21st birthday he had not only published various abstruse papers about mathematical relativity, but was also lecturing at the Department of Astronomy of the University of South Africa in Pretoria.

To list all the active amateur astronomers in South Africa would take many pages, and this is not the place to attempt it; but at least it is certain that enthusiasm and skill are increasing all the time. The amateurs, who study astronomy "for fun", are doing sterling work. Here, as in the professional field, the South African outlook is bright.

In this book we have done our best to give a comprehensive survey of the history of South African astronomy. It cannot be complete; much has been glossed over, and much has had to be left out, but we hope that what we have written will be of interest. It is, indeed, a great story. Tachard . . . La Caille . . . Herschel . . . Fallows . . . . Henderson . . . Maclear . . . Innes . . . Gill . . . The list of past famous astronomers is extensive, and today's successors are of equal lustre. Thanks partly to its climate, but mainly to its individuals, South Africa has been able to enrich the oldest of the sciences. Neither is the story at an end; we may be confident that the coming centuries will be as fruitful as those of the past.

INDEX

AD Leonis, 128
ADH Telescope 122, 124, 125
African Mythology, 19
Airy, Sir G., 45, 55, 59, 71
Airy Transit Circle, 70
Alcock, G. E. D., 149
Alden, H. D., 112
Aldrin, E., 137
Algieba, 128
Algol Stars, 149, 150
Alpha Centauri, 45-47, 61, 94, 101, 111
Alpha Crucis, 26
Alpha Orionis, 61
Amasi, 20
Andrews, A.D., 122, 128
Andrews, P., 119
Andromeda, 26
Ångström, A., 138
Ångströms, 138-140
Apollo 15, 133
Arc of the Meridian, 38-43, 55, 59, 64, 66, 68
Arequipa, *Peru*, 120-122, 125-127
Argo, 19, 116
Argo Navis, 61
Arizona Crater, 147
Armagh Observatory, 124
Armstrong, N., 137
Artificial Satellites, 104, 137
Asteroids, 98, 102, 106, 112
Astrographic Catalogue, 79
Astrographic Refractor, 79, 83
Astrography, 79
Astronauts, 133, 136
Astronomical Society of S.A., 106, 118, 152
Astrophysics, 144

Ba-Letsatsi, 20
Baade, W., 127
Bacher, J. M., 142
Back, E., 21
Backlund, *Russian Astronomer*, 96
Bailey, S. I., 121-122
Baker-Nunn Camera, 143
Bamberg Observatory, 142
Bauersfeld, W., 146
Bayer, J., 25-26
Bayly, *Astronomer*, 43
Beer, W., 133
Belgium, 142
Bennett, J., 150, 152
Bernard Price Institute of Geophysical Research, 146, 148
Bessel, F., 46-47
Bestbier, J. L., 37
Bester, M., 122, 128
Beta Orionis, 61
Beta Persei, 149
Biela's Comet, 45
Binaries, 93-94, 103-104
Birr Castle, 112
Black, I., 73
Bleksley, A., 146
Blink-Microscope, 101, 111
Block, D., 154
Board of Longitude, 43
Bok, B. J., 122
Bonn Dürchmusterung, 76
Bowler, T., 49
Boyden, U. A., 120
Boyden Observatory, 85, 120, 122, 124-129, 144
Bradley, J., 45-46, 55, 66
Breslau Observatory, 142
British Association for the Advancement of Science, 37
British Astronomical Association, 83
Bruce Astrograph, 122, 124
Brück, *Professor*, 125
Bruwer, J. A., 102, 106
Burbidge, M., 86
Bushmen, 20

Calton Hill Observatory, 73
Cambridge University, 114
Cannon, A., 120
Canopus., 61, 116
Cape, 91
  Longitude of, 25, 32, 39
Cape Astrographic Zone, 82
Cape Observatory *see* Royal Cape Observatory
Cape Photographic Catalogue, 82
Cape Photographic Dürchmusterung, 76, 79, 95
Cape Planetarium, 146
Cape Town
  Castle, 24, 27
  Grand Parade, 57
Carina, 61, 116
Carte Du Ciel, 79, 82
Cassegrain Design, 108
Cassini, C., 29
Cassini, G., 25, 29

Cassini, J., 29
Castor in Gemini, 26
Catchpole, R., 119
Cave Engravings., 19
Centaur, 26
Cepheids, 126-127, 150
Ceres, 49
Cherwell, *Lord.*, 115
Clark, A., 122, 125
Clark Telescope., 126
Clouds of Magellan., 141
Columbia University., 113
Comets., 149, 152
  1664, 21
  1665, 21
  1830, 45
  Biela's, 45
  Encke's 45, 70
  Halley's 62, 102, 114, 152
  1882, 75
  Gales, 93
  Ensor, 150
  Bennett, 152
  West's Comet, 152
Commission of Standards for Weights and Measures, 66
Concordia Gardens Observatory, 43
Cosmic Radio Waves, 140
Crab Nebula, 140
Craters, 147
Crux, 20
Crux Australis, 20
CSIR, 91, 104
Cycle of Celestial Objects, 49

Darwin, C., 15, 63
Deeming, T. J., 15
Deep Space Instrumentation Facility, 138
Deep Space Network, 138
De Kock, R. P., 150
De La Caille, N. L., 29-43, 92, 154
  Arc of the Meridian, 38, 43, 55, 59, 64, 66, 68
  Catalogue, 37
  Observatory, 37, 43, 57, 64
Delta Cephei, 126-127
De Sitter, W., 101, 107, 115
Dixon, J., 43
DQ Herculis, 144, 149
Dommanget, J., 122
Double Star Commission, 105
Double Stars, 93, 95, 99, 102-106, 130
Dreyer, J. L. E., 114
Dun Echt, 73-74
Dunsink Observatory, 125
Dutch East India Company, 23
Dwarf Nova System, 146
Dwarf Novae, 144-145
Dyson, F., 114-115

Earth Satellites, 137, 143, 152
Earthquakes, 102
Eddington, Prof. Sir A., 115
Elizabeth Telescope, 83, 85, 87
Elkins, *Astronomer*, 75
Encke's Comet, 45, 70
Ensor, G. E., 150
Eros, 82, 102
Escombe, H., 135
Eta Aquilae, 127
Eta Argûs, 61
Eta Carinae, 61-62
Evans, B. H., 15
Evans, D. S., 15, 84, 91, 115
Everest, Sir G., 43
Explorer 1, 137
Eyepiece Interferometer, 104

Fallows, F., 43-45, 51, 91-92, 154
  Observatory, 44-45
Feast, M. W., 119
Fennell, A. E., 92
Finley, W. H., 91
Finsen, W. S., 99, 102, 104, 106-109, 130, 137, 144
Flare Stars, 128
Flux Collector, 108-109
Forbes, A. F. I., 150
47 Tucanae, 128
France, 142
Franklin, J., 63
Franklin-Adams, J., 97-98
Franklin-Adams Photographic Maps, 98
Franklin-Adams Telescope, 102, 104, 107
French Academy of Sciences, 32

Gagarin, Y., 137
Galaxies, 50, 141
Gale, W. F., 93
Galileo, 25
Gamma Leonis, 128
Geary, T., 146
Gegenschein, 142
Geos I & II, 143
Geyer, E. H., 122
Gill, Sir D., 29, 55, 59, 71-79, 91-92, 95-96, 120, 135, 149, 154
  Telescope, 73
Goldfarb, S., 15
Goodricke, J., 126

Government Meteorological Observatory, 92
Government Trigonometrical Survey Office, 60
Grant, M., 136
Grazing Occultations, 152
Great Magellan, 17
Grootfontein, 147
Grove Primary School, 64

Hale Telescope, 125
Halley, E., 24
Halley's Comet, 62, 102, 114
Halm, J. K. E., 79, 91
Hamburg Schmidt, 124
Harding, G. A., 86, 91
Hartebeeshoek, 137-138, 140-141, 148
Hartebeespoort Outstation, 102, 104-105, 107-110
Harvard College Observatory, 120, 123
Harvard University, 122
Hay, W. T., 149
Heliometer, 73-75, 83
Hemming, J., 59
Henderson, T., 45-47, 91-92, 94, 154
Heninger, S., 143
Henize, K. G., 130
Hercules, 19
Herget, P., 105
Herodotus, 21
Herschel, Sir J., 15, 29, 31, 44, 50-53, 55, 57-58, 60-66, 92, 114, 133, 154
  Monument to, 64
Herschel, Sir W., 46, 50, 60, 127
  Catalogue, 51
Hervey, H. A., 44
Hoba West Meteorite, 147
Holden, F., 130-131
Holland, 107, 142
Holwarda, P., 126
Horrocks, H., 91
Hough, S. S., 79-80, 82, 91
Houghton, H. E., 150
Hubble, E., 127
Huggins, W., 120
Hussey, W. J., 130
Huygens, C., 107
Hydrogen Clouds, 130

Image Tubes, 123
Infra-red Photometry, 83
Innes, R. T. A., 92-99, 101-103, 106-107, 154
Institute of Chemistry, 134
International Astronomical Union, 105
International Education Board, 122

Jackson, C., 102, 113
Jackson, J., 82, 91
Jansky, K., 140
Jarrett, A. H., 17, 122-123, 125, 129, 131
Johnson, E. L., 102
Jones, Sir H. S., 80-82, 84, 91, 115
Juno, 74
Jupiter, 19, 25-26, 39, 45, 97, 137-138, 140, 148-149

Kappa Pavonis, 127
Khedivial Observatory, 114
Khonvum, 19
Kingzett, C. T., 134
Knipe, G. F., 105
Knox-Shaw, H., 114-115, 118
Kolbe, P., 26, 28, 33
  Observatory, 27
Krosick, *Baron*, 26

La Caille *see* De La Caille
Lamont, R. P., 130
Lamont-Hussey Observatory, 130-131, 142
Laser Transmitter, 143
Leavitt, H., 127
Le Bas, *Monsieur*, 25
Leiden Observatory, *Holland*, 101-102, 108
Leiden Observatory, *Johannesburg*, 102, 107-108
Leiden University, 107
Lesotho, 142
Lighthouses, 66
Lightwave, 138
Lindemann, F. A., 115
Lindsay, *Lord*, 73-74
Linnaeus, C., 133
Lindsay, E. M., 124-125, 129
Linné, 133
Livingstone, D., 66
Lohrmann, W., 133
London Science Research Council, 91, 104
Long-period Variables, 150
Lovell, B., 86
Lowell Observatory, 106, 147

Maclear, *Miss*, 74
Maclear, Sir T., 29, 31, 47, 49, 51-52, 55-59, 62, 64, 66-67, 70-71, 91, 154
McClean, F., 79
Magellanic Clouds, 26, 118, 124-127
Mann, W., 59, 67-70, 91
Mare Serenitatis, 133
Mariners, 138
Mars, 19, 45, 74, 104-105, 130, 137-138
Mars Patrol, 106
Mason, G., 43
Mason-Dixon Line, 43

Meadows, *Lieutenant*, 49
Men of the Sun, 20
Mercury, 19, 45, 137-138
Metcalf Refractor (Triplet), 122, 124, 126-128
Meteorites, 147
Meteorological Commission, 66
Meteorology, 92
Meteors, 152
Milky Way, 20, 26, 124, 140, 142
Mira, 128
Mira in Cetus, 126
Moon, 45, 73, 75, 114, 132-133, 135, 137-138, 147, 149, 152
Moonwatch Patrol, 152
Moon-worship, 19-20
Mount Brukkaros, Solar Observatory, 142
Muller, E. B., 39
Multiple Refractor Mounting, 87

Nandi, 19
NASA, 137-138, 140, 143
Natal Observatory, 132-136
Nebula, 61
Neison, E., 86
Neubabelsberg Observatory, 142
Nevill, E., 132
Nevill, M., 136
Newton, Sir I., 32
Newveld Mountains Observatory, 142
Nicolaysen, L. O., 148
Nicolson, G., 140
Nishimura Reflector, 124, 128
Nova Cygni, 144
Novae, 149

Occultations of Stars, 152
Octans, 45
Olifantsfontein, 143
Omega Centauri, 128
Orbiter Probes, 133, 138
Orion, 19
Orion's Belt, 20
Overbeek, D., 152
Oxford University, 114, 118

Palermo Observatory, 49
Papadopoulos, C., 152
Parallax Shifts, 111
Paraskevopoulos, J. S., 122
Pearse, M., 48
Period-Luminosity Law, 127
Perseus, 19
Petrie, Wm *junior*, 17
Photoelectric Photometers, 123
Photoelectric Photometry, 82-83, 105, 124
Photographic Star Atlas, 152
Photography, 51, 73, 75, 76, 82, 107, 111, 123-124, 152
Photometric Standards, 83
Photomultipliers, 108-109
Piazzi, G., 49
Pickering, E. C., 120
Pioneer 10, 140
Pioneer 11, 137
Planetaria, 146
Pleiades, 20
Ploughing Stars, 20
Poitevin, 39
Præsepe Cancra, 26
Precision Photoelectric Photometry, 83
Prentice, J. P. M., 149
Procyon, 61
Proxima Centauri, 47, 101, 111
Pygmies, 19

Quasars, 141

R R Centauri, 149
R R Lyrae Stars, 128
R R Pictoris, 144
Radcliffe Catalogue, 71
Radcliffe Observatory *Oxford*, 114
Radcliffe Observatory *Pretoria*, 82-83, 85, 88, 104, 114-119, 142, 152
  74 inch telescope, 117-118, 122
Radio Astronomy, 140
Radio Telescope, 138, 140, 148
Radio Wave, 138
Rambaut, A. A., 114
Rangers, 138
Rawson, *Rear-Admiral*, 137
Red Dwarfs, 128
Redman, R. O., 115
Reference Catalogue of Southern Double Stars, 95
Reid, W., 150
Rennie, J., 45
Republic Observatory, 85, 87, 92, 96-98, 102-106, 109, 113, 130, 142
  Leiden Station, 107-108
  Library, 105
Reunart, T., 96
Reunart Telescope, 96-97
Rhodes University, 148
Rhodesia, 142
Rigel, 61
Roberts, A. W., 149-150
Roberts, G., 106
Robinson, R., 124

Rockefeller Telescope, 102, 107, 122-124, 128
Ronald, W., 45
Ross, *Captain*, 64
Rosse, *Earl of*, 60, 112
Rossiter, R. A., 130
Royal Cape Observatory, 21, 29, 44-45, 51, 62, 64, 67, 70, 74, 79, 81, 83-85, 91-92, 104, 144
  Library, 91
Royal Geographical Society *London*, 142
Ryle, M., 86, 114

S S Cygni Stars, 145
St Helena Observatory, 24
Satellites, 138, 143
Saturn, 19, 29, 96, 107, 137, 149
Saunder, S. A., 114
Schlesinger, F., 111, 115
Schmidt, B., 125
Schmidt, J., 133
Schmidt, T., 122
Schmidt-Cassegrain Pattern, 125
Schmidt Telescope, 124-125
Schröter, J. H., 133
Science Research Council *London*, 118
Sea of Eternity, 19
Secchi, A., 120
Selenographical Society, 133
Selenography, 133
Shaka, 19
Shapley, H., 115, 121-122, 124-125, 127, 129
Sirius, 144
The Sky at Night, television series, 17
Slipher, E. C., 130
Smithsonian Institution, 142
Smyth, *Admiral*, 48
Smyth, C. P., 17, 49, 58-59, 91
Solar Constant, 142
Solar Eclipses, 19, 71
Solar Observatory, *Mount Brukkaros*, 142
Solar Radiation, 142
Sonnenberg Observatory, 142
South, J., 50-51
S.A. Association for the Advancement of Science, 92
South African Library, 66
South African Museum, 66
S.A. Literary & Scientific Institution, 63-64
South West Africa, 142, 147
Southern Cross, 20, 25-26, 45
Southern Hemisphere Observatory, 142
The Southern Sky, television series, 15, 17
Sputnik 1, 137
Stanley, Sir H. M., 67
Starry River, 20
Steavenson, W. H., 83, 114
Stellar Parallax Programme, 82
Stellar Parallaxes, 75
Stellar Photography, 107
Stellar Spectroscopy, 118, 120
Stepto, D., 148
Stevenson, D., 109
Stewart, M. B., 51
Stibbs, W. N., 118
Stone, E. J., 55, 71, 91
Stoy, R. H., 82-84, 86, 91
Sun, 47, 74, 101, 126, 134, 137, 140
Sun-worship, 19-20
Supernovae, 140, 152
Surveyors, 138
Sutherland Observatory, 21, 85-88, 91, 104-105, 110, 119-120, 144-145
  74 inch telescope, 122
Swastika, 20
Sweden, 142

Tachard, G., 20-21, 24-26, 92, 154
  Observatory, 22, 27, 142
Talbot, F., 51
Taurus, 140
Tea & Stars, 17, 63
Thackeray, A. D., 118-119
Three Healers, 20
Thuret, 25
Tide-Tables
  Natal, 135
Tortoise, 19
Tracking Stations, 138
Transit Circle, 79, 83
Transvaal Observatory *see* Republic Observatory
Tree of Life, 20
Trigometrical Parallax, 111
Tulbagh, R., 33, 39
Twin Telescope, 98

U Geminorum Stars, 145
Union Meteorological Department, 98
Union Observatory *see* Republic Observatory
Union Telescope, 112
University of Cape Town, 144, 146
  Department of Astronomy, 83
University of Michigan, 130
University of South Africa, 129
University of the Orange Free State, 129, 144
University of the Witwatersrand, 148
  Bernard Price Institute of Geophysical Research, 146, 148
  Library 15, 29
  Planetarium, 146

Yale Observatory, 111-113
Uranus, 50
UV Ceti, 128

Van Den Bos, W. H., 99, 101-104, 106-108, 130
Van Der Stel, S., 24
Van Gent, H., 102
Van Hoef, A., 122
Van Riebeeck, J., 24
Variable Stars, 95, 118, 124, 142, 145, 149-150, 152
Velghe, A. G., 122
Venus, 19, 43, 71, 74, 134, 137-138
Victoria Telescope, 79, 82-83
Vitamin C, 23
Von Braun, W., 137
Von Mädler, J., 133
Voyage de Siam des Peres Jesuites, 22, 24
Vredefort Ring, 146-148
Wales, 43
Walker, M. F., 144
Walraven, T., 17, 107-108, 110
Warner, B., 17, 118, 144-145, 154
West Germany, 142
White Dwarf, 144-146
Williams, *Lieut.*, 17, 63
Williams, E. G., 115
Windhoek, 142
Wood, H. E., 99, 100, 102, 106-107
Woolley, Sir R., 17, 21, 84, 86-87, 91, 104

Yale Observatory *Johannesburg*, 111-113, 142
Yale Observatory *U.S.A.*, 111
Yale Zone Catalogues, 111-112

Zeekoegat Observatory, 142-143
Zodiacal Light, 142
Zodiacal Stars, 135
Zulu Lore, 19